The Alchemists

Founders of Modern Chemistry

F. SHERWOOD TAYLOR

M.A., B.SC., PH.D.

Curator of the Museum of the History of Science, Oxford
Hon. Editor of AMBIX, the Journal of the Society for the
Study of Alchemy and Early Chemistry

ISBN 1-56459-002-X

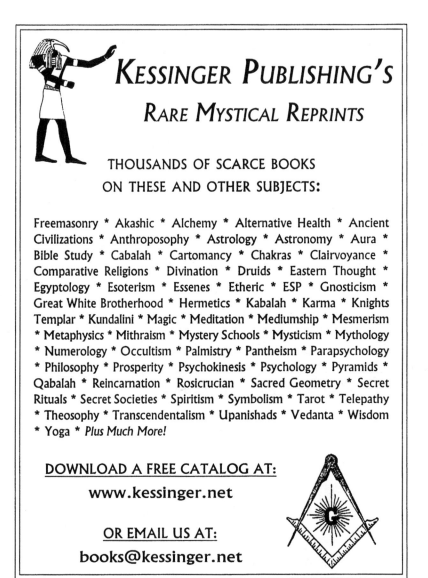

ALCHYMIA
(From Thurneysser's *Quinta Essentia*, 1570)

Contents

Illustrations

Illustrations

Preface

THE PURPOSE of this book is to give a short and clear account of the alchemists, their ways of thought and their contribution to man's achievement. This is a peculiarly difficult task, firstly, because we do not know all that the alchemists believed and did, owing to their deliberate and avowed concealment of the parts of their work that they considered most important; secondly, because the alchemical habit of thought was very unlike that of the modern reader; and lastly, because so much of their work remains unstudied and a great deal of what is known about them is still a battleground for controversialists. Where so much is in doubt the scholarly course would be to present the arguments for the various views; but since this would result in a long work, unreadable except by scholars, I have been content to give the views I have reached as a result of some twenty years of study, views which I do not regard as more than approximations to the truth, an interim report as it were.

If, as one of its historians has said, alchemy is the history of an error, why study it at all? There are, I think, three reasons for doing so.

In the first place, the hopeless pursuit of the practical transmutation of metals was responsible for almost the whole of the development of chemical technique before the middle of the seventeenth century, and further led to the discovery of many important materials. This is the commonly recognized contribution of alchemy, crystallized by John Donne in the words,

> And as no chemic yet th'elixir got,
> But glorifies his pregnant pot,
> If by the way to him befall
> Some odoriferous thing, or medicinal,
> *(Love's Alchemy)*

Secondly, we recognize today that science is not only a picture of the world here and now, but a human activity which can only be understood as a growing organism. Those who would understand the growth of chemistry must needs trace to their roots not only the fundamental chemical ideas, but the character and society of chemists; and it is in the alchemical laboratories that these took their origin.

Thirdly, the historian of today has a wider outlook than his predecessors and is not exclusively concerned with the political and economic aspects of past ages. The history of ideas is beginning to seem no less important than these, and the history of man's ideas concerning his relation to matter will, I am confident, appear to the historian of the future as a significant factor determining the history of the last five hundred years. The history of the attitude of man to matter in the years before 1600 has hardly been attempted, and it is hoped that this book will be a small contribution thereto.

F. SHERWOOD TAYLOR

Museum of the History of Science, Oxford

I

Introductory

FOR at least fifteen hundred years and perhaps longer there were, in the chief centers of civilization, a considerable number of men who carried out what we should call chemical operations with the ostensible purpose of changing base metals into gold. These men we call the alchemists, though the word came into existence long after the thing. We have all heard of alchemists, and most of us have a picture of them, somewhat confused with that of magicians or wizards. But this is a total error, for as far as we know the alchemists sought to accomplish their work by discovering and utilizing the laws of nature, and they never, or at least very rarely, attempted to bring about results by "magical" processes—by charms, spells, invocations of demons, etc.

They were not successful in discovering the laws that govern the changes in things and they did not go about their quest in the manner of modern science, but, for all that, the best type of alchemist was a quiet, lonely, and sincere seeker into the nature of things. The chief apparent purpose of alchemy, the making of gold, was, of course, so attractive to men that the way was opened to fraud; consequently there were, in the Middle Ages and probably earlier, a great number of charlatans who tricked men out of their money by faked demonstrations and so brought alchemy into disrepute. Our principal concern will be with the true alchemist rather than the fraud, of which our own times present more excellent examples for study.

Alchemy, we have said, flourished in the chief centers of civilization, and it followed the main stream of learning. It existed in China and India, but we cannot connect these Eastern alchemists with the main tradition which emerges in the Near East at a disputable date, not later than about 100 A.D. At that time alchemy was practiced in Alexandria and Egypt; it then spread through the Greek-speaking world. The Nestorians and Monophysites, exiled from Byzantium, were acquainted with the alchemical doctrines and carried these through Syria and Persia (450-700 A.D.), whence after the rise of Islam, came the learned men who translated the texts and gave them to the Arabic-speaking world.

The Arabs became enthusiastic alchemists. After 1100, Arabic texts were translated or paraphrased into Latin, and in the thirteenth and fourteenth centuries alchemy spread through the whole of Western Europe, where it continued to flourish exceedingly until, in the late seventeenth century, the rise of the modern scientific method made it ever less credible. None the less, a great many alchemical texts appeared in the eighteenth century and it was not until the dawn of the nineteenth century that the ancient tradition was broken.

The word tradition is truly applied to alchemy, because on the whole it looked backward, where modern science looks forward. The alchemist believed that the "ancients" knew the secrets of the work and could perform it, and

2

Introductory

his principal endeavor was to understand the meaning of their books. Modern science, on the other hand, looks forward to the time when her efforts will make known the things that have never been known. To those who made her she looks back with respect, with honor, but not with any belief that their works contain hidden secrets to be unraveled for her enlightenment.

And the alchemists that we are to study—they are also among the makers of modern science. In an age when men found it hard to interest themselves in things which had no human living interest, the alchemists set up the ideal of being able to conquer nature through natural processes. They sought not only to make gold, but *to perfect everything in its own nature*, and this is not far from the ideal of those who today would apply modern science as it should be applied. Like the modern scientist, they sought to do this by means of laboratory operations. The first laboratories we know of were alchemical laboratories. The alchemists were the first whom we know to have practiced distillation and sublimation, and they invented almost all the chemical apparatus that was in use up to the middle of the seventeenth century. If we had to assess their position in the history of science, we might best call them the Fathers of Laboratory Technique.

Yet if we turn to an alchemical treatise in the hope of appraising their achievement we find ourselves in a chaos. No literature is so maddeningly and deliberately obscure. The authors tell us that their books were written in such a way as deliberately to conceal the practice from all who had not been initiated into a certain secret which enabled them to understand. The apparatus is usually fairly clearly de-

3

scribed but the nature of the substance to be treated is concealed under cover-names. Thus Sol (the sun, the King) represents gold. But does it represent the metal that we call gold or some unknown entity the alchemists call "our gold"? Take the word "magnesia." Modern science gives this name to a well-known and definite substance, magnesium oxide, but the alchemists give the name to something that we cannot identify; nor were their contemporaries sure that they themselves could do so. Hear Chaucer, who was well acquainted with alchemy:—

> Also ther was a disciple of Plato
> That on a tyme seyde his maister to,
> As his book *Senior* wol bere witnesse,
> And this was his demande, in soothfastnesse,
> 'Telle me the name of the privy stoon.'
> And Plato answerde unto hym anoon,
> 'Take the stoon that Titanos [1] men name'
> 'Which is that?' quod he. 'Magnesia is the same.'
> Seyde Plato. 'Ye, sir, and is it thus?
> This is *ignotum per ignotius*.
> What is Magnesia, good sir, I yow preye?'
> 'It is a water that is maad, I seye,
> Of elementes foure,' quod Plato.
> 'Telle me the rote, good sir' quod he tho,
> 'Of that water, if it be your wille.'
> 'Nay, nay' quod Plato, 'certain, that I nille;
> The philosophres sworn were everichoon
> That they sholden discovere it unto noon,
> Ne in no book it wryte in no manere;
> For unto Crist it is so leef and dere,

[1] "Titanos" in the Greek alchemists means simply "lime"; "magnesia" is even then obscure and seems to have been applied to several different minerals.

4

Introductory

That he wol nat that it discovered be,
But where it lyketh to his deitee
Man for tenspyre,[2] and eek for to defende
Whom that hym liketh: lo, this is the ende,[3]

This concealment of the nature of materials is so general that only a very small minority of alchemical recipes can be interpreted in such a way that they could be repeated. Here is the foremost problem of alchemy. The alchemists were quite certainly performing real experiments with well-designed apparatus, but they rarely tell us what they put into the apparatus, and they describe effects which, as far as modern science can conjecture, could never have come to pass. Yet their works indicate that they were men of intelligence and seekers after truth.

The alchemists did not merely devote themselves to the attaining of a material purpose, the making of gold; for their works set out and develop a theory of natural philosophy, a view of the world, which was displaced by modern science. It was not without merit in its width and interpretation of experience, but it had not the power of science to predict physical phenomena, and it was based on premises that were, to say the least, doubtful. We cannot understand alchemy without the alchemical natural philosophy, which will take the reader into some strange countries of the mind.

[2] To inspire.
[3] Geoffrey Chaucer, "Canterbury tales." Chanouns Yemannes Tale, in *The complete works*, ed. W. W. Skeat. Oxford, 1894. 11. 1448-1471.

5

II

The Ideas of the Alchemists

THE ostensible object of the alchemists was to transmute other metals into gold. In the nineteenth century the metals were considered to be chemical elements, and chemical elements were considered as quite unalterable, except by combination. Transmutation was therefore supposed to be absurd. In the twentieth century we know it to be possible, in theory at least, but only by an expenditure of energy far beyond that at the disposal of the alchemist, who worked by the aid of the gentle fostering heats of the dung-bed or water bath. But before the time of Lavoisier even the word "element" did not exclude the possibility of transmutation, and so there was no theoretical reason to suppose the conversion of mercury into gold to be impossible. Before the mid-seventeenth century few were found to doubt that it could be done, though many doubted whether the alchemists had, in fact, succeeded in doing it.

In order to enter into the mind of the alchemist and show that there was reason in his odd procedures, we need to understand the science of his time. The Greeks were the initiators of theoretical science, and their conjectures and theories concerning the nature of matter went wherever al-

6

chemy went. Thus in Alexandria, in Byzantium, in Islam, and in Europe before the development of modern chemistry, the learned adopted the theories about matter and chemical change that had been held by the Greeks, especially by Aristotle and his commentators and by the Greek writers on medicine.

These theories are very different, indeed, from those of modern science, and their two chief doctrines were that of *matter and form*, and that of *spirit*. These three words have today entirely changed their meaning. Thus we say that sulphur and iron are different kinds of *matter;* but for Aristotle they were the same matter specified by different forms. When we talk of the form of a thing we mean its geometrical shape, but for the Aristotelians this was only one kind of form, and the form of a body was that which gave rise to all that we today call its "properties." *Spirit* today means either a volatile liquid or a courageous attitude or an incorporeal life; but the word *spiritus* or *pneuma* in ancient science meant literally *breath*, and could be applied to a vapor, a gas, a disembodied spirit or even to the Holy Ghost.

If we understand the ancient meaning of these terms and the use of them in ancient science we shall be much nearer to an understanding of alchemy.

In considering the ancient chemistry, one must forget most aspects of modern chemistry. No thought of atoms, of chemical elements, of pure substances, of conservation of mass, must enter our heads. We must think ourselves back to the position of the intelligent man viewing changes in things and changes in himself, and focusing his mind not so much on the details of the individual changes as on the idea of change.

The primitive conception of matter was an anthropo-

7

morphic one. Things are like ourselves. We are aware of a spiritual part and of a bodily part which is somehow controlled by it. It was natural, then, to analyze matter into a controller and controlled. We shall find this idea very frequently in the science of former times. The soundest and most penetrating analysis of matter along these lines was made by Aristotle in the fourth century B.C. There is an identity that persists during all change; by saying "iron *changes into* rust," we affirm a relation between iron and rust. It is reasonable to say that in such a change something changes and something persists. That which persists was called matter, that which changes was called form. We must not, however, think of Aristotle's "matter" as something that could exist by itself. In a statue, bronze was the matter and, let us say, Apollo was the form, but in bronze itself, earth and water were the matter and the "cause of the properties of bronze" was the form. Earth and water were accounted elements and were considered as an ultimate prime matter having the form of earth or water. So when Aristotle systematized this theory he assumed that in the last analysis there was only *one* ultimate matter which could take on an infinite number of forms, whence it follows that no material change is impossible, though some changes cannot take place directly. Aristotle recognized that many kinds of changes do not occur. A horse does not change into a lion or into a stone; yet it seemed that, when a horse died, it lost its form of horse and could be eaten by a lion, and so its matter could receive the form of lion. Alternatively, it could decay, thereby being resolved into less highly specified kinds of matter that could in turn receive the forms, e.g., of maggots. So the theory of matter and form seemed to indicate that if any substance could be reduced to a sufficiently

8

simple matter, this could be given the form of any other substance, so that, in theory, there was no reason to suppose that any substance was incapable of being changed into any other.

The earliest alchemists, who lived considerably later than the time of Aristotle, thought in terms of matter and form. So their endeavor to change, let us say, copper into gold, was planned as the removal of the form of copper (or, more picturesquely, as the death of the copper and its corruption), to be followed by the introduction of a new form, that of gold (which process was pictured as a resurrection).

But how was it to be done? The treatment of copper with certain solutions (especially solutions of sulphides) or the heating of it with sulphur made it lose its "metallic form" and become a black mass (of copper sulphide). This seemed to the alchemist to be the reduction of copper to matter without metallic form. But how was he to introduce the form of gold? That was the difficulty, and the theory of matter and form threw no light upon it. It was necessary to try to make a new complete being, gold, come into existence where no gold was before.

Where do we see this kind of thing in our daily experience? Almost everywhere. We see the generation of new animals from their parents, of plants from a seed, and of some creatures apparently from dead matter—for it was naturally enough supposed at that period that worms and flies and frogs and even creatures as highly organized as mice simply grew without parents out of decaying matter or mud, and it was a general belief that earth would put forth plants without the presence of seeds pre-existing therein. These simple beliefs of all primitive peoples, arising from observation uncontrolled by experiment, were systematized by the Greeks. The most obvious and important change of this kind was the coming-to-be of living organisms, and the first problem was to find reasons why new creatures with their complete organization (form) should arise where no creature was before.

In such a case it is notable that a new life has been produced: as God breathed the "breath of life" into the man of earth, so here the "breath of life" was entering into these creatures and organizing them. To do this was a very lofty function: that it was for the gods to make the crops grow was part of man's belief in the early days of Egypt, three thousand and more years before the Greeks began to think things out. And the stars had a hand in it, for crops grew according to the times of the year marked out by the march of the heavenly bodies. Yes, it seemed obvious that "breath of life" must come from the heavens and make new things come to be. What else is needed which does not come either from heaven or from the material to be converted into the new thing? Warmth was needed. The hen must sit on the egg, the child must grow in the warmth of the womb, the sun must warm the earth and make the river mud bubble and seethe with new life.

These are the primitive elements of the idea of generation: a seed, a soil, the breath of life from heaven and the gentle fostering warmth. These are the conditions that the alchemist set himself to imitate. He wants to make gold come to be, so he will grow it. He corrupts the other metals to form the soil, he can provide the gentle warmth in the dung-bed or water bath, but he requires the seed and the breath. Gold should not grow into gold, for a cabbage does not grow into cabbages, but into cabbage seed that will

10

grow cabbages. So we must have the seed of gold. Nothing else can grow into gold, for, as the alchemists are never tired of reminding us, of barley is bred barley, of a lion, a lion, and of gold, gold. How, then, was gold to be made fertile? On this problem the alchemists reflected deeply, and we shall discuss their answers in a later chapter.

Finally, how was the influence of the heavens to be brought to bear? The alchemist might work under the influence of the suitable heavenly bodies, making the chemical operations keep time with the planetary hours or with the seasons. This notion is found especially in the earlier alchemy, but it is less common than an attempt to obtain that influence in a form that could be handled and, in fact, used as a chemical reagent. This notion, to us extremely bizarre, is at the root of most of the alchemical practice and it can only be understood by those who have grasped what the ancients meant by "breath." The Greek *pneuma*, the Latin *spiritus*, the Indian *prana* have a very similar significance, but there is no word with such a meaning in modern English because the very idea has disappeared.

We know of matter, which we think of as without spiritual aspects; we know mind, which most of us who are not materialists regard as without material aspects; we do not know of anything, with qualities both of mind and matter, that could bridge the gap. But until the seventeenth century and even after, everybody recognized the existence of materials of different degrees of subtlety. There was gross matter that could be touched and handled—but even that, as we shall see, had a spirit in it; then there were clouds, smoke, vapor, exhalations, air, ether, the natural, vital, and animal spirits, the matter of spiritual beings; and it was only God himself who could be thought of as purely spiritual. There

11

were, in fact, thought to be materials of all grades of materiality and spirituality, approaching more nearly, in proportion as they were more subtle, to the nobler nature of mind.

Thus early Greek philosophers could say quite simply that the soul was air. They do not, of course, mean to imply that what we call a soul is the same as a mixture of oxygen and nitrogen: simply that the principle of life was a sort of breath. Some likewise regarded the atmosphere as a kind of reservoir of world soul, and thought that living beings existed by drawing in this air—a breath of life out of the soul of the world. This breath was believed to be distributed through the body by means of the lungs and blood vessels and to act as a number of subaltern minds, each situated in one of the bodily organs, controlling its action. Yet at the same time this breath could be transformed into *things*, and Aristotle, in a famous passage which is certainly one of the sources of the idea of alchemy, supposes that all the metals are made of two "exhalations," two vapors, one moist, one dry or smoky, that rise up through the earth. Here is the passage:

Some account has now been given of the effects of the secretion above the surface of the earth: we must go on to describe its operation below when it is shut up in the parts of the earth. Just as its twofold nature gives rise to various effects in the upper region, so here it causes two varieties of bodies. For we maintain there are two exhalations, one vaporous, the other smoky, and there correspond two kinds of bodies that originate in the earth, the "fossil" [1] and metals.

For the dry exhalation is that which by burning makes all the "fossil" bodies, such as the kind of stones that cannot be melted,

[1] The meaning of the word is simply "anything dug up." Until quite modern times the word simply meant mineral or stone.

realgar and ochre and ruddle and sulphur and other such things.[2] Most of the fossil bodies are colored ashes or a stone concreted from them such as cinnabar. The vaporous exhalation is the cause of all metals, fusible or ductile things, such as iron, copper, gold. For the vaporous exhalation, being shut in, makes all these things, and especially when in stones. By their dryness, being compressed and congealed into one thing, just like dew or hoarfrost, when it has been separated it generates these things. Hence these things are water in a sense, and in a sense not. For the matter was that of water potentially, but it is no longer, nor are they from water which has been changed through some affection, such as are juices. For copper and gold are not formed like that but each of them was formed by the exhalation congealing before water was formed. Wherefore all are affected by fire and have some earth; for they contain the dry exhalation. But gold alone is not affected by fire. This is the general theory of all these bodies but we must consider each of them in particular . . .[3]

These vapors, we notice, are so subtle that they can pass through stones, yet they can condense to form metals. Aristotle evidently considered the metals to be very closely akin, and the alchemists who followed him were thereby encouraged to think transmutation possible. The later alchemists identified the "dry vapor" with sulphur and the "moist vapor" with mercury; hence their theory that all metals are made of mercury and sulphur. In the earliest period of alchemy Aristotle's philosophy was not so popular as that of the Stoics and the Hermetics. But these schools relied even more than the Aristotelians on the notion of breath or spirit, which was for them at once the root and

[2] Realgar is arsenious sulphide, ochre and ruddle are clayey iron oxides; all three were used as red pigments.

[3] Aristotle, *Meteorologica.* Book III, Ch. 6 (378c).

the active principle of all things. So it was readily believed that not only could the metals be made of a "breath" coming up from the earth, but "breath" was also thought of as a power capable of directing events. We have seen that the heavens were thought to take a hand in the generation of new things, by calling forth the new form. Everyone believed that the planets continually influenced the formation of every new being, which is, of course, the theory that lies behind astrology. Here is a quotation from the *Bibliotheca Historica* of Diodorus Siculus (c. 50 B.C.):

> They (the Egyptians) say these gods (Isis and Osiris) in their natures do contribute much to the generation of all things, the one being of a hot and active nature, the other moist and cold, but both having something of the air; and that by these all things are both brought forth and nourished; and therefore that every particular being in the universe is perfected and completed by the sun and moon, whose qualities, as before declared are five: a spirit (breath) or quickening efficacy, heat or fire, dryness or earth, moisture or water, and air, of which the world does consist as a man is made up of head, hands, feet and other parts . . . And therefore they called the spirit Zeus which is such by interpretation, because a quickening influence is derived from this into all living creatures, as from the original principle; and upon that account he is esteemed the common parent of all things.[4]

We see here evidence that, even before the time when we first hear of alchemy, it was believed that the sun and moon are agents in making new things and that the means of bringing about these changes is a living spirit or breath contained in the creatures of earth.

[4] Diodorus the Sicilian, *The historical library*. In fifteen books. Translated by G. Booth. London, 1700, p. 4.

The Ideas of the Alchemists

Now the influence of the sun, moon and planets upon this quickening spirit in the terrestrial bodies had to be accounted for. What medium was there by which, for example, the sun and moon, the stars and planets could effect the germination of a plant, as they certainly appeared to do? This medium was believed to be this same breath or spirit, which was thought to be the original emanation from God, which animated dead matter, an idea which persisted throughout the whole history of alchemy. To quote again, this time from Synesius, later Bishop of Ptolemais, (c. 400 A.D.):

> For then it was the Good
> Source of the spirit of man
> Was divided without division;
> And immortal mind, efflux
> Of divine parents,
> Descended into matter
> Scanty indeed, but whole and one everywhere,
> The whole diffused into the whole
> Revolved the vast hollow of the heavens
> Preserving all this whole.
> It is distributed into different forms
> Part of it in the courses of the stars,
> Part of it the choirs of angels,
> Part likewise in the heavy bondage
> Found an earthly form,
> And disjoined from its parents
> Drank dark oblivion, blind in its cares
> Wondering at the joyless earth.[5]

These ideas concerning "spirit" and its function in nature were very familiar at the time when we first hear of al-

[5] Synesius Episcopus, "Hymnus I." (Migne, *Patrologia Graeca*, Paris, 1859, Vol. 66, col. 1589.)

chemy. The Stoic philosophy was then in its ascendant. The Stoics held that all things were "body," material in the sense of taking up space. They conceived of all the changes in the world as the result of changes in body, achieved by the working of the primal fire, which brought into action the seed-like potentialities of things and caused them to develop in accordance with the plan inherent in their nature. The agent in effecting all such changes was a "breath," *pneuma*. The Hermetic philosophy, which was also prevalent in the earliest period of alchemy, maintained very similar views concerning the universality and efficacy of spirit. So the idea of a seed in things, brought to development by warmth and activated by "breath," was not some ancient primitive notion revived, but the latest pronouncement of the most approved philosophy and science of the day.

To sum up all this, we find that the philosophers and men of science, at the time when alchemy first appears, thought of the changes of the natural world as a drama in which this subtle matter—spirit or breath—played the principal part. We shall often have occasion to allude to it, and as the English words breath and spirit have been given other meanings today we shall normally give it its Greek name of *pneuma*.[6]

It may at once be said that alchemy still remains an unsolved problem, but that some reason can be discovered in

[6] The words spirit and soul have, of course, theological meanings, but these were only slowly defined and rendered precise. The alchemists speak of both spirit and soul in things; in Greek these are *pneuma* and *psyche*, in Latin *spiritus* and *anima*. Spirit and soul are regarded as male and female; the soul must, moreover, be the soul *of* some body, and is separated as a whole and returns as a whole, whereas spirit is not necessarily the spirit of something, but an entity that can be subdivided and of which there can be less or more.

the alchemical writings if there are borne in mind the three
ideas set out in this chapter:

1) The theoretical possibility of transforming any kind of
   matter into any other.
2) The need for such a transformation to take place by the
   corruption of the material to be transformed and the
   generation of a new form therein.
3) The power of a subtle but not wholly immaterial being,
   pneuma, to become a metal, to aid and direct generation,
   and to evoke new forms.

I I I

The Origin of Alchemical Practice

IT is interesting to note that although the classical Greeks had theoretical ideas about the origin of metals and the nature of change, we have no reason to suppose that they practiced either chemistry or alchemy. It may be that the stimulus of a partial apparent success in transmutation was needed before the science or art of transmuting metals was thought to be worth pursuing. It seems pretty certain that before there was any theorizing on the matter, practical technicians had been preparing white metals resembling silver and yellow metals resembling gold. It is difficult to know just how ancient this practice was. Campbell Thompson believed that a fragmentary Assyrian tablet (seventh century B.C.) referred to the making of "silver," but the alchemists themselves supposed their art to have come from Egypt. Thus the alchemist Zosimus, writing, it must be remembered, about 300 A.D. when Egyptian science and mythology were no longer a living tradition, begins one of his books thus:

Herein is established the book of the Truth.
Zosimus to Theosebeia greetings!
The whole of the kingdom of Egypt, lady, depends on these

18

two arts, that of seasonable things [1] and that of minerals. For that which is called the divine art, whether in its dogmatic and philosophic aspect or its phenomena in general, was given to its wardens for their support; and not only this art, but also those which are called the four liberal arts and the technical manipulations, for their creative capacity is the property of kings. So that, if the kings permitted it, one who had received the knowledge as an inheritance from his ancestors would interpret it, whether from oral tradition or from the inscribed columns. [2] But he who had the knowledge of these things in full did not himself practice the Art, for he would have been punished. In the same way, under the Egyptian kings the workers of the chemical operations and those who had the knowledge of the procedure (?) did not work for themselves, but served the Egyptian kings, working to fill their treasuries. For they had special masters set over them and a strict supervision was kept, not only upon the chemical operations, but also upon the gold-mines. For if anyone in mining found anything, it was a law among the Egyptians that it should be handed in for entry in the public register. [3]

It is an undoubted fact that the winning and working of gold were in ancient Egypt the subject of a priestly craft centered upon the temple of the god Ptah at Memphis. The god was "master of gold smelters and goldsmiths," his temple the "goldsmithy" and his priests were distinguished by such titles as "Great Wielder of the Hammer," "He who knows the Secret of the Goldsmiths." The same alchemist Zosimus, writing about 300 A.D., tells us that "I have examined in detail a furnace in the ancient temple of Mem-

[1] Possibly a reference to astronomy or astrology.
[2] Located in the temples; known as steles.
[3] "First book of the completion," §1. (Berthelot, *Collection des anciens alchimistes grecs*, texte grec, p. 239.)

phis . . ."; [4] and the context implies that this was at least similar to those employed by alchemists.

From Egypt, too, we hear of hints of the making of substitutes for gold. Gold plating was known and also gilding with gold leaf, and in the Roman period (after 30 B.C.) we hear of fire-gilding by means of mercury. The Egyptians practiced the art of coloring gold with varnishes and by corrosive liquids—both of which arts we meet in the works of the first alchemists. There is some evidence, too, that they "increased the weight" of gold by debasing it with other metals—a practice described by some of the earliest alchemists.

Egypt is a land where damp and its concomitant decay have no power. Buried in its sands, in tombs, in mummy cases and ruins, are found written papyri. Papyrus was the earliest kind of paper, made by sticking together slips of the bark of the papyrus reed. It was an article of common use—little less so than is paper today. In Roman Egypt it was used for books, legal documents, letters—even for wrapping-paper. Vast numbers of papyri have been unearthed and have thrown a flood of light on the life and habits of the period when they were written and, incidentally, give us some information concerning the attempt to make precious metals.

Two extremely interesting papyri were dug up more than a century ago and are referred to as the papyri of Leyden and Stockholm. [5] The authors of these are unknown, but

[4] "On apparatus and furnaces," §1. (Berthelot, *op. cit.*, p. 224.)

[5] (a) *Papyri Graeci Musei Antiquarii Publici Lugduni Batavi,* ed. C. Leemans. Leyden, 1885.

(b) *Papyrus Graecus Holmiensis,* ed. O. Lagercrantz. Upsala, 1913. Berthelot gives a translation of the chemical parts of the papyrus of Leyden in his *Introduction à l'étude de la chimie des anciens et du moyen âge.* Paris, 1889, pp. 28ff.

the format and handwriting indicate that they were written towards the end of the third century A.D. They contain some hundreds of recipes for the preparation (or falsification) of gold, silver, asemos,[6] precious stones, and dyestuffs. It is interesting that these should be lumped together in a single treatise, and it is clear that the coloring of a metal so as to imitate gold or silver, or of glass to imitate a precious stone, was thought of as quite analogous to the dyeing of a piece of cloth.

How did the authors of these papyri try to make gold and silver? Here is a recipe from the Papyrus of Leyden:

56. Asemos one *stater*[7] or copper of Cyprus 3 *staters*: 4 *staters* of gold; melt them together.

In other words, turn 24-carat gold[8] into 19-carat or 10-carat gold. This type of recipe is common enough. It seems that it was thought of not as a mere mixing, say, of gold and copper, but as an increase of the quantity of gold at the expense of its quality. Here is a less crude recipe from the same papyrus:

87. To increase the weight of gold, melt it with a fourth part of cadmia. It will become heavier and harder.

Cadmia was an impure mixture of oxides of base metals, copper, zinc, arsenic, etc., obtained from copper smelter's flues. The effect of the process would be to smelt these

[6] A white silver-like metal. The word in modern Greek simply means "silver"; in the works of the alchemists it seems to mean a "white silver-like metal."

[7] Measure of weight.

[8] Egyptian gold was not always refined, so that the original gold might contain silver and copper and the final product might be even more debased than the recipe indicates.

oxides to metal which would mix with, debase, and augment the weight of the gold.

These papyri contain a great variety of other recipes for gold-making. "Gold" is made not only by debasing genuine gold as described above, but also by surface treatments. Thus objects of base gold are heated to redness with iron sulphate, alum and salt. These evolve sulphuric and hydrochloric acids which remove the base metals from the surface of the gold, leaving a thin layer of pure gold which, after polishing, makes the whole object look like pure gold. Other recipes describe gilding.

An interesting and primitive recipe, also from the Papyrus of Leyden, runs:

To give objects of copper the appearance of gold so that neither the feel nor rubbing on the touchstone [9] will discover it; particularly useful for making a fine-looking ring. This is the method. Grind gold and lead to a dust fine as flour: two parts of lead for one of gold, then mix them and incorporate them with gum, coat the ring with this mixture and heat. This is repeated several times until the object has taken the color. It is difficult to discover because the rubbing (i.e., on the touchstone) gives the mark of an object of gold and the heat consumes the lead [10] and not the gold.

Gilding with an amalgam of mercury and gold in the modern style is also explained. A number of recipes mention colored gums or varnishes or dye liquors to tint metals superficially in the style of a lacquer, and numerous methods

[9] A hard black stone on which the gold was rubbed, leaving a bright metallic streak. The color and extent of the streak enables an expert to judge the purity of the gold.

[10] I.e., oxidizes it to litharge which melts and runs off.

22

of making gold-colored paints or inks with various yellow lacquers and pigments are given.

Much attention is also paid to the making of silver and "asemos," a white alloy resembling silver. Here is a recipe for making silver: [11]

Take copper which has been prepared for use and dip it in dyer's vinegar and alum and leave it to soak for three days. Then melt one *mina* [12] of the copper, some Chian Earth and Cappadocian salt and flaky alum up to six drachmae. Smelt it carefully and it will be excellent. Add not more than 20 drachmae of good and tested silver which will make the whole mixture imperishable (untarnishing).

The process is, first, a superficial cleaning of copper (the mixture of alum and vinegar is very effective). Next the copper is melted with a sort of fuller's earth, with salt and with "flaky alum," a term which in the works of the alchemists seems to be used in some places for a composition containing arsenic. A fusion, carefully performed in order not to drive off all the arsenic, gives a white or whitish yellow copper-arsenic alloy. By fusing this with silver a brilliant white alloy, containing perhaps 77% copper, 19% silver, and 3% arsenic, would be obtained. If the "flaky alum" is simply alum, then a very base silver would result.

We find in these papyri very clear evidence that attempts to make gold and silver, sometimes genuine and sometimes fraudulent, were being carried out in Egypt before 300 A.D. We should say that these papyri were the work of alchemists, were it not that their gold-making is treated as an entirely matter-of-fact and practical process. There is no

[11] *Papyrus Graecus Holmiensis* (1st recipe).
[12] 1 mina = about 1 lb. = 100 drachmae.

theory or philosophy of chemical change in them, nor are there any hints of revelations from gods or traditions from ancient philosophers, nor any concealing of methods under symbols, nor rhapsodies about the divine character of their Art. None the less, these papyri are the earliest documents[13] which reveal the idea of making precious metals, and the methods they use, moreover, are very like those of one of the groups of early alchemists.

We cannot regard these actual papyri as the source from which true alchemy developed; for they are not as old as some of the alchemical texts. This is shown by the fact that one of them mentions the alchemist Democritus. But they give us the valuable information that practical goldsmiths were trying to make gold and silver in Egypt not long after the time when the first alchemists were writing. It is a reasonable inference that these papyri reveal to us an ancient tradition of Egyptian metal-working, and that this tradition contributed to the earliest alchemy.

[13] The papyri are much older than the existent alchemical *manuscripts;* but the first alchemical writers lived perhaps two centuries before the papyri of Leyden and Stockholm were written.

IV

The First Alchemists

THE first group of alchemists of whom we have any
knowledge lived in the Greek-speaking parts of the
world at a period which cannot be exactly defined.
It is generally thought to begin somewhere in the great
period of Greek science which commenced about 300 B.C.
in Alexandria and had much declined by about 200 A.D. All
that remains of them are a number of manuscripts which
contain the disordered fragments of a few of their works.
The oldest manuscript is not older than about 1000 A.D.
But just as we should know by the style, language, and sen-
timents whether a piece of verse was written in the sixteenth,
seventeenth, or eighteenth century, so we can hazard guesses
at the time at which these works were composed. Unfortunately we can find no references to alchemy in the works
of writers whom we can date until near 500 A.D. However,
we know that Zosimus mentions the temple of Serapis (at
Alexandria) which was destroyed in 390 A.D., and that
he therefore lived before that date, and that Zosimus regarded "Democritus" and Mary the Jewess as "ancient"
authors.

According to usual opinion, the evidence indicates that
the earliest of these works were written about 100 A.D., but

some scholars would put the alchemical writings attributed to "Democritus" back to about 250 B.C.

The fact that the earliest alchemists are not mentioned by their non-alchemical contemporaries suggests that, for the first two or three centuries of its life, alchemy was a semi-secret though written tradition, pursued by a few obscure persons in the vast city of Alexandria, the home of strange branches of learning and the meeting-place of Eastern, Western, and old Egyptian creeds and practices.

Who were these alchemists? Their identity, like almost everything in the study of this subject, is obscure. There are names of authors at the head of the treatises, but they are nearly all obviously false attributions. We have the names of fifteen so-called ancient alchemists. Of these, nine are obvious false attributions:

Democritus (Greek philosopher: b. about 470 B.C.)
Isis (Egyptian goddess)
Iamblichus (Neoplatonic philosopher: d. about 330 A.D.)
Moses (Hebrew prophet)
Ostanes (Legendary Persian sage)
Cleopatra (Egyptian queen)
Hermes (A god or legendary sage)
Agathodaemon (Phoenician serpent-deity)
Pibechios (A god = Apollo Bechis)

We need not prove this in the case of gods and goddesses; but the falseness of the attributions in the other case is as patent as would be the falseness of the attribution of a work on organic chemistry to Shakespeare. In many cases the names appear to have been added long after the work was written, probably in order to enhance the value of the manuscript.

There are five names which may not be pseudonyms:

The First Alchemists

Komarios, Mary the Jewess, Chymes, Petasios, and Pammenes; and one of these, Mary the Jewess, seems to have been a real person and a great discoverer in practical science (pp. 38-40).

Later than these was Zosimus of Panopolis, who, at a date not far from 300 A.D., wrote an encyclopedia of alchemy, of which parts survive. We feel that we can attach a certain personality to the author of the work attributed to "Democritus," to Mary the Jewess, and to Zosimus; but the rest are just so many names attached to fragments of text. The later Greek alchemists who wrote after 400 A.D. are no more than commentators who tried to explain what the older alchemists intended and, it would seem, knew but little more about it than we know today.

To the modern reader it will seem curious that the alchemists so often assumed as *noms de plume* names more famous than their own. Greek alchemists, writing about 200 A.D., put at the head of their treatises the names of mythological characters such as Hermes, Isis, Agathodaemon, of famous philosophers who had lived many centuries earlier, such as Leucippus or Democritus or even Moses, of kings and queens such as Cleopatra or Cheops—attributions which would find a parallel if Darwin, without any attempt to modify his style, had published the *Origin of Species* as a lost work of Francis Bacon, St. Thomas Aquinas, Queen Anne, or Edward the Confessor. The reason for such attributions was, in all probability, the enormous respect of the ancients for the still more ancient, and their belief that the world was regressing from a state of wisdom and goodness to that of folly and impiety. Those who took such a view of history naturally respected an old book more than a new one, and a manuscript with a fine old name at its head was much more valuable than one whose author appeared to be some unknown contemporary of the reader.

The alchemist, unlike the chemist, did not seek to advance his art by discovery of new methods, but by the rediscovery and new interpretation of older writers whom he believed to have possessed the secret. Consequently he wished his books to appear to be ancient. This tendency persisted to a diminishing extent throughout the history of alchemy. Ramón Lull, St. Thomas Aquinas, Roger Bacon, and other famous medieval philosophers had spurious treatises on alchemy laid at their doors, sometimes within a few years of their deaths. These older false attributions are rarely puzzling. The alchemist to whose works the name of Moses is attached makes no effort to assume the character, and writes exactly like the author whose text is assigned to Iamblichus, a Neoplatonic philosopher who lived a couple of thousand years later than the Hebrew prophet.

In the Middle Ages, however, the false texts were often written by followers of the great man, who imitate his style; but even so we are not often in doubt. It is well to remember, however, that throughout our study of the earlier alchemy we have always to question the authorship of every text. None the less, the fact that most of these Greek texts are pseudonymous does not make them less interesting. They were, at any rate, written by the first alchemists, and we can learn from them something about the kind of people their authors were, and about the work they were doing.

We can say, first of all, that, although the earlier alchemists wrote in Greek, they were not Greeks, but in all probability Egyptian or Jewish. They were not Christians, for they talked in terms of the mythology of Egypt—of Isis, Horus, Hermes (= Thoth). They were familiar with the

28

names and ideas of Greek philosophy, and they were, at
the same time, practical laboratory workers. Some were
women. Omitting Cleopatra, whom we do not suppose to
have concerned herself with such matters, we have the
names of Mary the Jewess, of Paphnutia, and of Theosebeia,
the sister of Zosimus. There was evidently some exchange
of information among them. A fragment of a letter from
Zosimus to his sister runs:

In the same way your priest Nilus moved me to laughter, burning his lead-copper alloy in a baker's oven,[1] as if he was baking
bread, burning it with cobathia[2] for a whole day. Blinded in
his bodily eyes he did not realize his method was bad, but he
blew up the fire and after cooling and taking out his product,
showed you cinders. Being asked where the whitening was, he
was at a loss and said that it had penetrated into the interior.
Then he put in copper and colored the cinder, for meeting
nothing solid it passed out and disappeared into the interior,
the same being true for the whitening of magnesia. Hearing this
from his opponents, Paphnutia was much laughed at, and you
will be laughed at, too, if you do the same. Greet Nilus, the
cobathia-burner.[3]

We have no more than a vague notion of what Zosimus
is talking about (probably an attempt to give a silvery appearance to an alloy of lead and copper by means of arsenical vapors from some mineral), but the passage gives us a
tiny picture of the society of the chemical operators in
Egypt.

What were these earliest alchemists trying to do? They
were one and all concerned with the artificial making of

[1] An earthen pot in which bread was placed, covered with hot embers.
[2] Probably some arsenical and sulphurous mineral.
[3] Berthelot, *Collection des anciens alchimistes grecs*, texte grec, p. 191.

some precious material, usually gold and silver, but sometimes precious stones or the famous Tyrian purple dye of the ancients. We need here concern ourselves only with their attempts to transmute metals. We have already seen that this was reasonably thought to be possible.

The men of those times had no conception that there existed one and only one exactly-defined chemical individual called *gold*. There were all sorts of golds, some very good, others not so good. They were, however, all "gold" to the ancients, not mixtures of one pure gold with varying proportions of base metal. Gold was something shining, heavy, yellow, untarnishable, and resistant to fire.

How would the alchemists know whether their final product was in fact silver or gold, for they had, of course, no possibility of chemical analysis? We know of two tests that were actually applied, the test of the touchstone and the test by fire. The gold was rubbed on a hard black stone and its quality was judged from the color and extent of the yellow streak produced. The professional gold- and silversmiths, moreover, would have had the expert's delicacy of sense that would have made them suspicious of anything that did not look or feel quite right. Next in importance was the test by fire. Pure gold, however long it is heated, remains unchanged. This test would rule out alloys chiefly composed of base metals, but a slight oxidation at a high temperature was evidently not considered incompatible with gold. Modern jeweler's gold will not stand prolonged heating without change, since it always contains copper. Most native gold is likewise contaminated with copper, and this would help to minimize the failure of the artificially produced gold to satisfy the conditions of the fire-test.

A third possibility was the measurement of the specific

gravity of the metal. The high specific gravity of gold cannot be imitated by any alloy of the metals known to the ancients, and the practical man certainly would reject any piece of gold that felt too light. But we must doubt whether any numerical measurements of specific gravity were made, for although such measurements had been used by Archimedes to detect impurities in gold, there is little evidence that this test was applied in the early days of alchemy.

Thus for an alchemist to believe that he had prepared gold, he would have had to make a metal which closely resembled gold in color and hardness, which was of high density, and which was little affected by atmospheric action.

It is evident that it was much easier to produce a colorable imitation of silver than of gold, for there are a good many white alloys of about the same density as silver, but there are very few yellow alloys and all of them are much less dense than gold.

The early alchemists tried methods of preparing white and yellow alloys by fusion and also by coloring the surfaces of metals. They also tried more elaborate methods involving the use of distilled substances. The first two methods are much easier to comprehend, but it was the last that was the source of most of later alchemical and chemical technique.

The principle that lay behind the simplest alchemical processes seems to have been the attempt to introduce the properties which the base metal lacked. It seemed to the alchemist that a metal might be rendered white or yellow by removing the property of yellowness or whiteness from another substance and introducing it into the metal. The color was a sort of *activity* and therefore a *pneuma* or "spirit." They tell us that "a tingeing *pneuma* gives its color

31

to metals"; that the color of plants is their *pneuma*. So we find that the early alchemists generally made use of yellow and white substances in their efforts to make the yellow and white metals.

In most cases this principle led to nothing, but we find certain apparent successes, coincidences of color between reagent and product which appeared to support this rule. Chief of these were the whitening of copper by white arsenic, and the yellow coloration imparted to copper and silver by weak solutions of the yellow polysulphides obtained by boiling lime and sulphur, or even by solutions of yellow dyestuffs. In case the reader may doubt if anyone would try to imitate gold by dyeing a white metal, I may cite the line from John Donne

> And like vile lying stones in saffron'd tin
>
> *(Elegy* VIII, l. 13)

Let us first consider the case of silver. The recipes by which the authors of the technical papyri made this have already been discussed (p. 23). In the true alchemical texts, the few recipes for making it follow similar lines. Thus in some cases white alloys of various metals were prepared. One interesting recipe indicates that if copper is "whitened," its alloy with silver will not have a dark color. This indicates the preparation of an alloy of copper, silver, and arsenic. In a late recipe, "Making of Silver with Tutia," an alloy is prepared in which silver, lead, zinc, and copper are all present.

A method which was much more frequently described was the attempt to whiten copper by means of arsenic. Yellow arsenic, i.e., arsenious sulphide, is found as the fine yellow mineral *orpiment*, which was used by the ancients as a pigment. This substance was very familiar to the alchemists,

32

who knew how to "whiten" it by subliming it in presence of air, a process, of course, which oxidized it to the white arsenious oxide to which we give the name of "arsenic" today.

If copper was cleansed by boiling it with alum and acid liquids and was then melted with some arsenic compound, mixtures or alloys of copper and copper arsenide resulted which were white, lustrous, and much like silver in appearance. Alternatively, the arsenic compounds might be smeared on the copper, which was then heated; whereupon a superficial layer of the white copper-arsenic alloy was formed. This whitening of copper was, it seems, not regarded by all of the alchemists as a true making of silver, but it was a shining example of the possibility of altering the color of a metal, and this whitened copper was sometimes used as a first stage in the attempt to make gold.

The making of gold was a process in which it was much more difficult to obtain even an appearance of success. As far as we can interpret the recipes, the earliest alchemists employed four methods:

1) The making of yellow alloys of base metals, much like brass.
2) The preparation of debased gold.
3) The superficial coloring of metals or alloys.
4) A series of very complex processes in which distilled liquids were employed or in which metals were subjected to the action of vapors.

The first three of these methods have some resemblance to the old technical methods that appear in the papyri. Moreover, they have been in use in modern times for making artificial jewelry. The fourth method, though very obscure, is more important for our purpose as being the ancestor of the processes of the later alchemists.

Brass-like alloys, including some of the alloys of copper, tin, and zinc, used in modern times under the names of *ormolu, oroide, Mannheim gold*, etc., were certainly prepared by the Greek alchemists. Zinc was not known to the ancients in metallic form, and these brassy alloys were prepared by smelting mixtures of the other metals or their ores with *cadmia*, which was a mixture of metallic oxides containing a variable proportion of zinc, found as a deposit in the flues of smelting furnaces. It cannot have been easy to get reproducible results with this impure and variable material, which may account for the complexity of the recipes for its use. These recipes are by no means easy to understand, but it seems that the Greek alchemists prepared a number of brassy alloys containing copper as chief constituent together with tin, lead, zinc, iron, silver, mercury, or some of these. Though we must doubt if any goldsmith ever took any of these to be gold, their yellow color must have given hope that success was near.

The most nearly successful process was that of doubling gold, already mentioned on page 21. This type of recipe for making gold employs a considerable quantity of the precious metal and is called by the Greek alchemists *diplosis* or "doubling," i.e., a doubling of the weight of gold. It depends mostly on the fact that, while silver gives a greenish and copper a reddish color to gold, the admixture of both copper and silver hardly alters its tint. We need not suppose that the alchemist who melted up gold with silver and copper necessarily regarded himself as in any way falsifying gold; for he probably believed that the gold acted as a seed which,

34

nourished by the copper and silver, grew at their expense until the whole mass became gold. In such recipes are described the preparation of alloys of many types, some of which are today legalized on the European Continent, just as are 18-carat gold and other gold-copper alloys in Great Britain. They include

1) Gold-copper alloys with small quantities of other metals, notably zinc and arsenic. This corresponds to our modern 14-18-carat gold, possibly made somewhat lighter in color by the presence of zinc.

2) Gold-copper-silver alloys, similar to the above, but reproducing the color of pure gold more closely.

3) Alloys containing much copper and some silver and gold. The yellow color of these derives chiefly from the copper, and the addition of precious metal probably prevents the alloy from tarnishing readily.

Here is an example to show how the alchemists set out their recipes:

DOUBLING ACCORDING TO MOSES

Copper of Calaïs, one ounce, orpiment, native sulphur, one ounce and native lead one ounce: decomposed realgar (arsenic sulphide) one ounce. Boil in oil of radish, with lead, for three days. Put it in a roasting pan and place this on the coals, till the sulphur is driven off, then take it off and you will find your product. Of this copper take one part and three parts of gold. Melt it, fusing strongly and you will find it all changed to gold, by the help of God.[4]

The text of the recipe is corrupt, but the product would contain about 66% of gold, 33% of an alloy of copper, lead, and arsenic, and, in color and resistance to chemical

[4] *Ibid.*, p. 38.

35

action, would closely resemble pure gold. The alchemist who had carried out such a recipe might reasonably think that the gold had converted the lead and copper into its own substance with the help of the golden-yellow color from the orpiment, and there is evidence in the technical papyri that their authors thought in this way. But if gold could convert something like its own weight of copper and silver into gold, why should it not convert greater, nay very large, quantities of base metal?

It was not unreasonable, then, to suppose that gold could act as a seed or a ferment growing in and transforming a mass of base metal, as leaven transforms dough. Actually, of course, while the "doubling" recipe gives a gold-like metal, a small proportion of gold will hardly affect the properties of copper or other base metals, and the use of a little gold as a seed or ferment could not have been successful in producing a yellow metal. Thus, in a text, probably of the third or fourth century,[5] we read that "this water (the 'divine water,' pp. 44-45) acts like leaven, producing the like by the like. As the leaven of bread, being a little, ferments a great quantity of dough, so also a little gold can ferment the whole dry matter." It appears that if gold was to be made from metals and the "divine water," a leaf of gold was dissolved in the divine water beforehand, "for barley engenders barley and the lion, a lion, and gold, gold." No information is given about the practical procedure. The "divine water" has such a wide meaning that it is not possible to discover what kind of process was intended, and we can only discover the idea of using a "ferment."

Yet another process used by the Greek alchemists was the superficial coloring of metals. These superficial treat-

[5] *Ibid.,* p. 145, §3.

ments were hardly regarded as a true making of gold, and as a rule the word "dyeing" and not "making" is used to describe them. These methods also find their counterpart in modern practices. Then, as now, three chief methods of coloring metals were employed:

1) Coating the metal with a tinted lacquer composed of gums, etc., as brass is treated today.
2) Tinting the metal with solutions which form a thin superficial layer of sulphides.
3) Treating debased gold by removing the base metal from the surface by corrosive substances, such as the sulphur trioxide derived from the calcination of the sulphates of iron and copper, so leaving a layer of fairly pure gold on the surface. At the present day, nitric acid is used instead of the sulphates.

The methods that have hitherto been described are intelligible processes by which something resembling gold or silver could be made, but such methods play but a small part in alchemy as a whole. The typical alchemical process involves volatile substances ("spirits") and is conducted by means of distillations and sublimations. All pictures of alchemical apparatus or laboratories show us apparatus for handling volatile substances. As far as we know, the process of distillation was invented by the earliest alchemists, who used a great deal of complex and well-designed apparatus for preparing volatile substances and treating metals with their vapors.

Nothing that can really be called distillation was known before the time of the alchemists. It seems that a sort of sublimation of liquids was occasionally practiced. Thus sea water was heated in covered cauldrons and the drops con-

37

densed on the lids were shaken off and used as drinking water; again "oil of pitch" was made by heating pitch and condensing the vapor on fleeces. Mercury was made by heating cinnabar on an iron saucer in a pan covered by a pot called an "ambix," on which the vapor of mercury condensed (Fig. 1), but none of these pieces of apparatus could

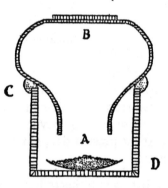

Fig. 1.—Probable appearance of the mercury-still of Dioscorides. (Courtesy of *Annals of Science*.)

be called a still. A still or alembic consists of three parts, a vessel in which the material to be distilled is heated, a cool part to condense the vapor, and a receiver. The traditional form of still is, indeed, that shown in Figure 2. A is usually called the still-head, B, the body and C, the receiver—though many other terms were used.

This piece of apparatus was invented by the Greek alchemists, or, at least, is first described in their writings, and it continued to be listed in chemical catalogues as late as 1860! It is first described by Mary the Jewess, though we do not know that she invented it, and her description is quoted by Zosimus:

38

The First Alchemists

I shall describe to you the tribikos. For so is named the apparatus constructed from copper and described by Mary, the transmitter of the art. For she says as follows:

Make three tubes of ductile copper a little thicker than that of a pastry-cook's copper frying-pan: their length should be about a cubit and a half. Make three such tubes and also make a wide

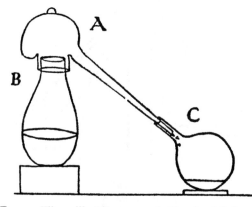

Fig. 2.—The still. (Courtesy of *Annals of Science.*)

tube of a handsbreadth width and an opening proportioned to that of the still-head. The three tubes should have their openings adapted like a nail to the neck of a light receiver, so that they have the thumb-tube and the two finger-tubes joined laterally on either hand. Towards the bottom of the still-head are three holes adjusted to the tubes, and when these are fitted they are soldered in place, the one above receiving the vapor in a different fashion. Then setting the still-head upon the earthenware pan containing the sulphur, and luting the joints with flour paste, place at the ends of the tubes glass flasks, large and strong so that they may not break with the heat coming from the water in the middle. Here is the figure.[6]

[6] *Ibid.*, p. 60.

39

The Alchemists

Figure 3 accompanies it, but we know it was drawn perhaps seven hundred years after Zosimus wrote, and if we consider what the text says and what the Greek words mean we come to something like Figure 4.

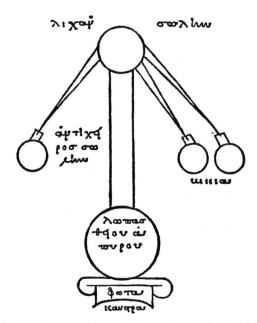

Fig. 3.—Mary's three-armed still. (Courtesy of *Annals of Science*.)

The standard type of still is described by an alchemist called Synesius, who is commenting on a book of Democritus (now lost), who apparently was the first to describe it:

What he (Democritus) says, O Dioscorus, is as follows . . . 'And put it into a flask on the hot-ash-bed, not having the fire direct, but on a gentle hot-ash-bed, which is a *kerotakis* (p. 46). During the action of the heat, there is adapted to the

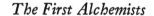

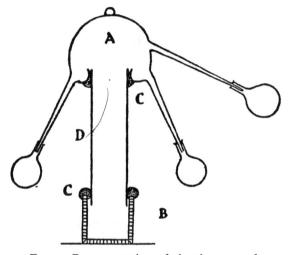

Fig. 4.—Reconstruction of the three-armed
still. (Courtesy of *Annals of Science*.)

flask above, a glass apparatus having a *mastarion* (a breast-shaped cup) fitting on to it. And put it on top of it and receive the water which comes up through the breast and keep it and putrefy it. This is called divine water." [7]

Figure 5 is appended in the manuscript and there seems no reason to doubt its accuracy.

Zosimus describes yet another kind of still which remained popular up to the eighteenth century and was termed a cold-still, because the liquid in the body was not boiled but only gently warmed. This is not illustrated in the manuscript but may be reconstructed as in Figure 6.

The discovery of distillation can be summed up as in Figure 7. First we have (a) the simple condensing of the vapor from sea water on a potlid, as described by Alexander

[7] *Ibid.*, p. 236.

41

Fig. 5.—The still of De-
mocritus. (Courtesy of
Annals of Science.)

Aphrodisias, a commentator on Aristotle; then (b) the condensation of mercury in a flask-like vessel, as described by
Dioscorides. We may conjecture that the next stage (c)
was the turning in of the edge of the lid to make a container
for the distillate; then (d) the addition of a pipe to lead it
off. Such a still would give trouble if the liquid were boiled

Fig. 6.—The cold-still of Zosimus.
(Courtesy of *Annals of Science.*)

at all quickly, because it would boil up into the still-head, so we may suppose that Mary put the wide vertical tube between this and the boiling pan (e). Democritus attained the same end by using a long-necked flask (f) while Zosimus, who did not wish to boil the liquid, kept the old type.

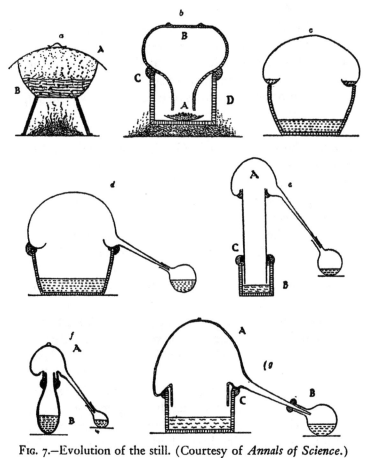

FIG. 7.—Evolution of the still. (Courtesy of *Annals of Science*.)

43

These stills are well designed for distilling moderately volatile liquids, such as water, and we naturally ask what the alchemists distilled. Here we find another difficulty. Every chemist will agree that no apparatus could be worse than these for the distillation of sulphur, which boils at $444°C$ and whose vapor condenses to a liquid which solidifies on cooling. Yet the substance put in the body of the still is usually described as *theion apuron*, literally "sulphur without fire." This is a recognized term in non-alchemical authors for "native sulphur." But the accounts of the distillations and the apparatus are such that it seems quite impossible to believe that the substance which we call sulphur is what they distilled.

We know that the product of distillation was called *theion hudor*. The word *hudor* in Greek means water, and *theion* means either sulphurous or divine, a coincidence of meaning in which our alchemists doubtless rejoiced. The descriptions of the processes certainly indicate that the product of distillation was a "water" and not a solid like sulphur. There were, according to the texts, numerous different "sulphurs," and we do not know what these alchemists intended by the word.

But there are two recipes which tell us to put something other than "sulphur" in the body of the still, and in each case the material in question is eggs. The distilled liquid was collected in three fractions; first, a clear distillate called "rain-water"; then a pale golden liquid called "oil of radish"; then a dark yellowish green liquid called "castor-oil." Now, if actual eggs are distilled, we obtain, first, a large quantity of clear liquid, faintly alkaline; then, a golden yellow somewhat oily-looking distillate containing ammonium sulphide, ammonia and pyridine bases; lastly, a very dark yellowish

44

thick liquid containing pyridine bases and tarry products. This corresponds pretty closely to the alchemist's description, and the products do what he tells us they should do. Thus the second distillate renders arsenic yellow, as he informs us, presumably by reason of the sulphides it contains.

Why should the alchemists wish to distil eggs? I suppose that they were seeking to extract the "breath of life," the *pneuma,* that was present in the egg, which of all things has the most obvious potency of generation. Moreover, the yolk of the egg had a promising golden color, and as we have seen, the sulphurous liquids obtained by distilling eggs had a yellow color and could confer it on certain materials of the Art, such as white arsenic and silver.

. We find in the Greek alchemical texts long lists of "waters" which may be solutions or distilled products, and it seems likely that all manner of plant and animal products were distilled by the Greek alchemists, as they were later by the Arab alchemists. I would suggest, then, that *theion apuron,* "native sulphur," was a general term for something from which *theion hudor,* "a divine (or sulphurous) water," could be distilled, and that this meant a sort of water that had the power of acting upon metals, corroding or coloring them. The name "divine water" was evidently applied to the yellow calcium polysulphide solution made by boiling lime, sulphur, and water; also to mercury, and to yellow dye-liquors used for superficial tinting. Zosimus, in fact, uses the word as a generic term for all liquids useful in the Art.

So, as far as we can see, distillation was invented simply as a means of obtaining a liquid useful for attacking or coloring metals, and we certainly hear nothing of the use of stills for any non-alchemical purposes until some seven hundred

45

years after their first use in alchemy, when we hear of them in books of workshop recipes.

The Greek alchemists name about eighty different pieces of apparatus. Furnaces, lamps, water baths, ash baths, dung-beds, reverberatory furnaces, scorifying pans, crucibles, dishes, beakers, jars, flasks, phials, pestles and mortars, filters, strainers, ladles, stirring rods, stills, sublimatories, all make their first appearance as laboratory apparatus in their works and have persisted in somewhat modified forms to the present day.

In addition to these, they had curious reflux apparatus designed for treating metals with vapors. The most important of these was called a *kerotakis*, which was the name applied by the ancients to the artist's palette. The artists of the time painted with a mixture of pigment and melted wax, and their colors had to be kept hot during use on this *kerotakis*, which was a metal sheet, shaped like a bricklayer's trowel, and kept hot over a pot of charcoal. The alchemists probably had the intention of softening the metals and impregnating them with color in the same way as the wax of the artist was softened and mixed with pigment. The actual palette, a metal plate of triangular or rectangular form, was soon reinforced by the addition of other pieces of apparatus.

The first stage in the evolution of the apparatus was its adaptation to the treatment of metals by means of heated vapors. A vessel below the palette itself contained a vaporizable substance capable of attacking metals, while an inverted cup placed above the palette condensed it to liquid which flowed back. The nearest modern analogy to the developed *kerotakis* apparatus is the reflux extractor. Further developments of the *kerotakis* were in the direction of

46

elaborating the heating and condensing arrangements and of providing a form of grating or strainer, perhaps to prevent large solid fragments of metal from falling into the base.

There is no clear explanation of how the apparatus was to be used, but the following is a possible explanation. Sulphur, sometimes mixed with arsenic sulphides, was placed in the lower part of the apparatus (Fig. 8) and on the *kerotakis* (P) were placed the several metals to be treated: copper, lead, perhaps gold and silver also. The condensing covers were then luted into position, a small hole being provided to allow escape of the heated air. This was covered by a little cup. The fire was then started; the vapor of sulphur attacked the metal and the sulphide which was formed dissolved in or mixed with the excess of liquid sulphur and ran through the sieve or grating into the base or "Hades." [8] The black mixture of sulphur and sulphides remaining there was the "scoria" or "black lead." This was desulphurized by heating or by treatment with lime or "oil of nitre" and then smelted. The resulting metal was, of course, an alloy of the metals originally used, but probably also contained some sulphur and some arsenic (if arsenic was used in the attacking mixture). Cadmia (p. 21) or arsenic ("the etesian stone") seems to have been introduced at some stage in the process. In the subsequent gentle roasting and smelting, the cadmia presumably added zinc to the alloy and so produced a kind of brass or latten containing copper, lead, and zinc. The alloy so obtained was then sometimes employed in the "doubling" of gold.

The process outlined seems excessively complex for the mere preparation of an alloy, but it must be remembered

[8] Cf. p. 59.

47

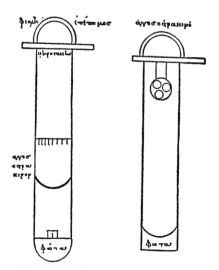

Fig. 8.—The *kerotakis* or reflux apparatus. Above, as shown in the Greek manuscript; below, a conjectural restoration (M = metals, P = palette). (Courtesy of *Journal of Hellenic Studies*.)

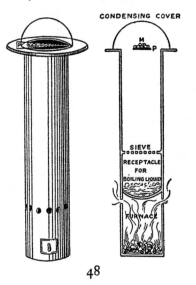

48

that these alchemists had no means of discovering that they were simply making an alloy, and no means of finding out the composition of what they had made. They were trying by empirical means, guided by an incorrect theory, to color metals, and in case of success they would not know which of the substances or processes had contributed thereto. Almost all ancient recipes, whether in alchemy or workshop practice, contain and preserve for centuries apparently useless materials and procedures which could not be discovered to be useless without scientific tests. The treatment of metals by the *kerotakis* is probably a very ancient one, deriving from Egyptian and Jewish sources, and (as is seen in the quotations on pages 58 and 62, which almost certainly refer to it) was regarded with a certain mystical reverence. It was not only a chemical preparation but to some extent a symbolic rite.

Some of the pieces of apparatus were much more complicated than those shown in the figures, but the principle seems to have been the same. The processes carried on in the *kerotakis* apparatus were said to involve a continuous and successive "blackening, whitening, and yellowing," followed sometimes by *iosis*. The latter word is of doubtful significance; it may mean "empurpling"—imparting the color of a violet (*ion*), but may only mean "removal of rust or tarnish" (*ios*).

This phenomenon is a difficult one to account for. The conversion of copper and the other metals used into their black sulphides accounts for the blackening; and the smelting to a yellow metal, for the yellowing. The whitening is very much less easy to explain. If, as is probable, the black product was dried before smelting, it might be whitened as a result of the efflorescence of salts derived from the "divine water." Alternatively, some white material such as compounds of mercury, arsenic, or antimony may have been added in order to bring about the desired whiteness. *Iosis* is probably merely a final tinting or perhaps a cleaning of the metal produced.

We cannot, in fact, interpret these complicated and fragmentary recipes well enough to say definitely what occurred, but it is clear that they were felt to be enormously significant by the alchemists who studied them, and that they gave rise to impressive symbolic writings.

V

The Earliest Alchemical
Signs and Symbols

THE chemical symbol and formula is familiar to everyone. We write H for an atom of hydrogen, K for an atom of potassium, H_2O for a molecule consisting of two atoms of hydrogen and one of oxygen. This is at once a convenient shorthand and a means of expressing the composition and structure of compounds. This type of representation goes back to the earliest period of alchemy, for the Leyden papyrus (c. 250 A.D.) contains the signs for gold and silver, and the page of figures known as the *Gold-making of Cleopatra* (Fig. 9), which is probably as old as anything in alchemy, contains the symbols for gold, silver, and mercury, and perhaps also for the lead-copper alloy and for arsenic. We have considerable lists of the signs in the oldest Greek manuscripts. Some of them are derived from the signs of the planet with which the metals were associated, others from pictorial representations of the things symbolized, others from the initial letters of the name.

The connection of the planets and metals is certainly ancient, and it persists throughout the whole of alchemy.

The metals have all received planetary signs. Gold re-

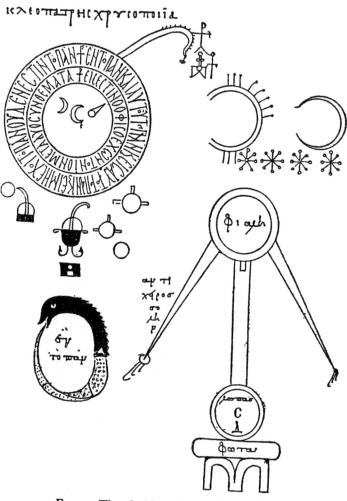

FIG. 9.—The *Gold-making of Cleopatra*.
(Courtesy of *Journal of Hellenic Studies*.)

ceived the sign ⊙ , representing the sun; silver, the sign of the waxing moon ☾ ; mercury, that of the waning moon ☽ (Hermes speaks of "that which drips from the waning moon"); copper has the sign of Venus (Aphrodite—Isis—Hathor) ♀ ; lead has the sign of Saturn ♄ ; iron has the sign of Mars ♂ . There remain the signs of electrum and tin. Tin has in these old lists the symbol of Hermes ☿, and electrum, that of Zeus ♃ . In later times (between 500 and 700 A.D.) the symbol of Hermes was given to mercury in place of that of the waning moon. Electrum was no longer considered a separate metal, and its symbol was then given to tin.

This system of metals and planets remained unaltered throughout the subsequent development of alchemy and indeed until Dalton suggested his new chemical symbols based on the atomic theory. But the form of some of the signs was altered, and the forms found in printed works are:

Gold ⊙		Lead ♄	
Silver ☾		Iron ♂	
Copper ♀		Tin ♃	
	Mercury ☿		

The system could be extended to express the composition of alloys by putting together the symbols of the metals composing them very much in the modern style. Thus the seven signs reproduced in Figure 10 represent:

1) Gold. (Shown as the sun with a single ray.)
2) Gold filings.
3) Gold leaf.
4) Calcined gold.
5) Electrum. (Sign of gold and silver combined.)
6) Chrysocolla. Solder of gold. (Two gold signs joined.)
7) *Malagma* of gold. (Mixture of gold.)

53

The notation for the alloy electrum, arrived at by combining the signs of gold and silver, its constituents, contains the germ of modern chemical symbolism, though of course the distinction between mixtures and compounds was not yet made.

The connection of gold with the sun is well known and has been the subject of discussion by Elliott Smith and

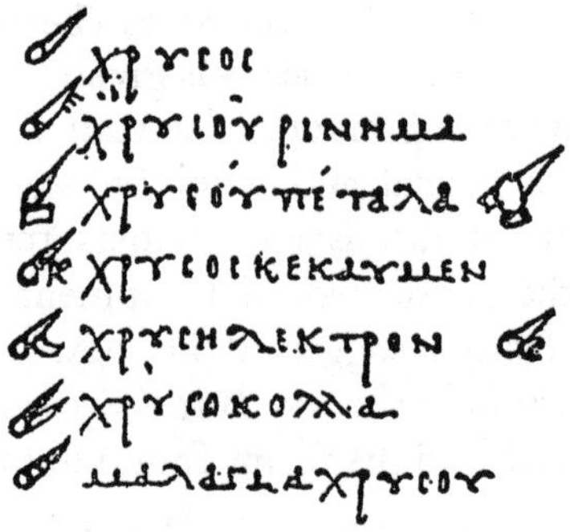

Fig. 10.—Combinations of signs to express modifications of gold.

others. It does not appear a distant step from this to connect silver, or "white gold" as the Egyptians termed it, with the silvery moon. The connection between copper and Aphrodite (Isis–Hathor) does not seem so clear. Perhaps the connection of Aphrodite with Cyprus, the source of copper (*chalkos kuprios*) is the origin of the association. Ares or Mars is associated with iron, obviously as the result of their common connection with warfare. The Assyrians and Babylonians named iron Ninip after their god of war. Lead was associated with Osiris by the Egyptians, according to the alchemical texts, and it receives the sign of the planet Saturn. Osiris does not truly correspond with Saturn or Kronos, though the motive of dismemberment is found in the myths of both. The idea of lead as a heavy metal being connected with the slowest moving planet is a possible explanation of the connection with Saturn. Osiris, on the other hand, in Egyptian myth represents water and liquidity in general, and the fusibility of lead possibly supplies the connection. The Assyrians and Babylonians named lead Anu, after a sky god who had some resemblance to Saturn. The connection of tin with Hermes seems difficult to explain and also that of Zeus with electrum. It is possible that there was never any very strong connection, the association being needed in order to fit the seven metals to the seven planets. When mercury had to be fitted to the scheme, it was obviously to be associated with Hermes, or Mercury, on account of its mobility and "subtlety."

Another explanation of the association is believed to go back to the Sabæans, who inherited much of the starlore of the Assyrians, namely that the color of the metal was associated with the color of the planet. The sun was golden, the moon silvery, Saturn leaden, . . . but it is difficult to fit the red-color of Mars to iron, unless the color of its rust was considered, nor the brilliant blue white lustre of Venus to copper. This association of metals and planets is not only alchemical, for it is mentioned by Celsus (quoted by Origen, *Contra Celsum*, VI. xxii) about 180 A.D., who gives it a Persian origin:

These things are obscurely hinted at in the accounts of the Persians, and especially in the mysteries of Mithras, which are celebrated amongst them. For in the latter there is a representation of the two heavenly revolutions—of the movement, viz., of the fixed stars, and of that which takes place among the planets, and of the passage of the soul through these.

The representation is of the following nature: There is a ladder with lofty gates, and on the top of it an eighth gate. The first gate consists of lead, the second of tin, the third of copper, the fourth of iron, the fifth of a mixture of metals, the sixth of silver, and the seventh of gold. The first gate they assign to Saturn, indicating by the lead the slowness of this star; the second to Venus, comparing her to the splendor and softness of tin; the third to Jupiter, being firm and solid; the fourth to Mercury, for both Mercury and iron are fit to endure all things, and are money-making and laborious; the fifth to Mars, because, being composed of a mixture of metals, it is varied and unequal; the sixth, of silver, to the Moon; the seventh, of gold, to the Sun,—thus imitating the different colors of the two latter.

It is evident, then, that there was no general agreement in antiquity concerning the metals to be associated with each planet.

The representation, e.g., of the metal mercury by ☿, the sign of the planet Mercury, was not a mere sign and nothing more, as is the representation of an atom of oxygen by the letter O, for it conveys the idea which lies behind this association, that the heavenly motions of this planet were causally connected with the terrestrial activities of that metal. This notion frequently appears in alchemical texts, though it does not seem that there was commonly any attempt to accommodate the time of alchemical operations to favorable conjunctions of the heavens.

But in alchemical works of every period we find quite a different kind of symbolism, designed to display the meaning of the operation to the instructed while concealing the practice from the ignorant. The changes in the chemicals contained in the alchemical vessels made a deep impression on the minds of those who saw them. The bright metal became a black formless mass, a stinking corruption; then another process brought this dead mass back to the state of metal again, and, so it seemed to them, perhaps because they wished it so, a more glorious and excellent metal. The process was, in fact, a symbol of what the age was seeking, what was found alike in Christianity and the mystery-religions—death and resurrection. In this life you must die to sin and be born again; moreover, the body, base metal now, will die and corrupt to blackness, but will be raised from its corruption new, glorious, and incorruptible like gold. This analogy runs through alchemy from its earliest times, and the more mystically-minded alchemists seem to have regarded this aspect as the most important part of it. It seems, indeed, that some authors took the actual physical appearances as a symbol of a more universal process of death and regeneration, while the more chemically inclined took death and regeneration as a symbolic expression of the chemical process.

Of this kind of symbolism more will be said in Chapter XI, but its character can be best expressed by quoting two famous allegorical passages, one from *The Dialogue of Cleopatra and the Philosophers*, the other from the works of Zosimus.

The passage from the *Dialogue of Cleopatra and the Philosophers* is considered to be among the earliest alchemical writings, probably of the second century A.D. The work is only a fragment, and of this only a part can be here quoted, but it will convey to the reader both the difficulty of interpreting the alchemical texts and their singular impressiveness. There also exists a page of symbolical drawings called the *Gold-making of Cleopatra*, which is reproduced as Figure 9. It seems to illustrate her principal themes, the unity of all things, and death and revivification through a "water."

The *Gold-making of Cleopatra* indicates these notions briefly. It consists simply of a page of symbolic drawings. In the center of the Serpent Ouroboros who eats his tail are the words ἕν τὸ πᾶν—"One is all." Another emblem contains the symbols of gold, silver, and mercury enclosed in two concentric circles, within which appear the words *One is the serpent which has its poison according to two composi-tions* and *One is All and through it is All and by it is All and if you have not All, All is Nothing.* A distillation apparatus is clearly figured, and also other alchemical apparatus and symbols not clearly understood.

The Dialogue is too long to be quoted in full, but the passages that follow will illustrate its character.

. . . Then Cleopatra said to the philosophers. "Look at the nature of plants, whence they come. For some come down from the mountains and grow out of the earth, and some grow up from the valleys and some come from the plains. But look how they develop, for it is at certain seasons and days that you must gather them, and you take them from the islands of the sea, and from the most lofty place. And look at the air which ministers to them and the nourishment circling around them, that they perish not nor die. Look at the divine water which gives them drink and the air that governs them after they have been given a body in a single being."

Ostanes and those with him answered Cleopatra. "In thee is concealed a strange and terrible mystery. Enlighten us, casting your light upon the elements. Tell us how the highest descends to the lowest and how the lowest rises to the highest, and how that which is in the midst approaches the highest and is united to it, and what is the element which accomplishes these things. And tell us how the blessed waters visit the corpses lying in

Hades fettered and afflicted in darkness and how the medicine of Life reaches them and rouses them as if wakened by their possessors from sleep; and how the new waters, both brought forth on the bier and coming after the light penetrate them at the beginning of their prostration and how a cloud supports them and how the cloud supporting the waters rises from the sea."

And the philosophers, considering what had been revealed to them, rejoiced.

Cleopatra said to them. "The waters, when they come, awake the bodies and the spirits which are imprisoned and weak. For they again undergo oppression and are enclosed in Hades, and yet in a little while they grow and rise up and put on divers glorious colors like the flowers in springtime and the spring itself rejoices and is glad at the beauty that they wear.

For I tell this to you who are wise: when you take plants, elements, and stones from their places, they appear to you to be mature. But they are not mature until the fire has tested them. When they are clothed in the glory from the fire and the shining color thereof, then rather will appear their hidden glory, their sought-for beauty, being transformed to the divine state of fusion. For they are nourished in the fire and the embryo grows little by little nourished in its mother's womb, and when the appointed month approaches is not restrained from issuing forth. Such is the procedure of this worthy art. The waves and surges one after another in Hades wound them in the tomb where they lie. When the tomb is opened they issue from Hades as the babe from the womb." [1]

This is a mysterious way of describing some alchemical operations and materials, and it appears to refer to the type of process that has already been described on pages 46-49. The author, by writing in this style, conceals the real nature

[1] Berthelot, *Collection des anciens alchimistes grecs*, texte grec, pp. 289-299.

of the process and must be thought to be writing for those who already know it. Why, then, write at all? The explanation is, I think, that such a work is in reality a kind of poem, expressing the wonderful analogies that the author sees between the great world with its seasons, and growth and death and regeneration, and the process of the alchemical work. It is, in fact, religious and technical, a sort of rejoicing in the wonderful phenomena of chemical change, and at the same time renders them more wonderful by assimilating them to the great events of nature which find their unfailing response in the heart of men. Indeed the paragraph that follows this quotation compares the philosopher contemplating his work to the mother contemplating the fruit of her womb, and the chemical waters are compared to her milk. The symbolism in its most developed form becomes an allegory, the chemical process being seen in terms of a parallel human story.

The most complete examples are the Visions of Zosimus, contained in his treatise *Of Virtue*.

LESSON 1

1. The composition of waters, the movement, growth, removal, and restitution of corporeal nature, the separation of the spirit from the body, and the fixation of the spirit on the body are not due to foreign natures, but to one single nature reacting on itself, a single species, such as the hard bodies of metals and the moist juices of plants.

And in this system, single and of many colors, is comprised a research, multiple and varied, subordinated to lunar influences and to the measure of time, which rule the end and the increase according to which the nature transforms itself.

2. Saying these things I went to sleep, and I saw a sacrificing

60

priest standing before me at the top of an altar in the form of
a bowl. This altar had 15 steps leading up to it. Then the priest
stood up and I heard a voice from above saying to me, "I have
accomplished the descent of the 15 steps of darkness and the
ascent of the steps of light and it is he who sacrifices, that re-
news me, casting away the coarseness of the body; and being
consecrated priest by necessity, I become a spirit." And having
heard the voice of him who stood on the bowl-shaped altar, I
questioned him, wishing to find out who he was. He answered
me in a weak voice, saying, "I am Ion, the priest of the sanctuary, and I have survived intolerable violence. For one came
headlong in the morning, dismembering me with a sword, and
tearing me asunder according to the rigor of harmony. And
flaying my head with the sword which he held fast, he mingled
my bones with my flesh and burned them in the fire of the
treatment, until I learnt by the transformation of the body to
become a spirit."

And while yet he spoke these words to me, and I forced him
to speak of it, his eyes became as blood and he vomited up all
his flesh. And I saw him as a mutilated little image of a man,
tearing himself with his own teeth and falling away.

And being afraid I awoke and thought, "Is this not the situation of the waters?" I believed that I had understood it well,
and I fell asleep anew. And I saw the same altar in the form of a
bowl and at the top the water bubbling, and many people in it
endlessly. And there was no one outside the altar whom I could
ask. I then went up towards the altar to view the spectacle.
And I saw a little man, a barber, whitened by years, who said
to me "What are you looking at?" I answered him that I mar-
velled at the boiling of the water and the men, burnt yet living.
And he answered me saying, "It is the place of the exercise
called preserving (embalming). For those men who wish to
obtain virtue come hither and become spirits, fleeing from the
body." Therefore I said to him "Are you a spirit?" And he
answered and said, "A spirit and a guardian of spirits."

61

The Alchemists

And while he told us these things, and while the boiling increased and the people wailed, I saw a man of copper having in his hand a writing tablet of lead. And he spoke aloud, looking at the tablet, "I counsel those under punishment to calm themselves, and each to take in his hand a leaden writing tablet and to write with their own hands. I counsel them to keep their faces upwards and their mouths open until your (*sic*) grapes be grown." The act followed the word and the master of the house said to me, "You have seen. You have stretched your neck on high and you have seen what is done." And I said that I saw, and I said to myself, "This man of copper you have seen is the sacrificing priest and the sacrifice, and he that vomited out his own flesh. And authority over this water and the men under punishment was given to him."

And having had this vision I awoke again and I said to myself "What is the occasion of this vision? Is not this the white and yellow water, boiling, divine (sulphurous)?" And I found that I understood it well. And I said that it was fair to speak and fair to listen, and fair to give and fair to receive, and fair to be poor and fair to be rich. For how does the nature learn to give and to receive?

The copper man gives and the watery stone receives; the metal gives and the plant receives; the stars give and the flowers receive; the sky gives and the earth receives; the thunderclaps give the fire that darts from them. For all things are interwoven and separate afresh, and all things are mingled and all things combine, all things are mixed and all unmixed, all things are moistened and all things dried and all things flower and blossom in the altar shaped like a bowl. For each, it is by method, by measure and weight of the 4 elements, that the interlacing and dissociation of all is accomplished. No bond can be made without method. It is a natural method, breathing in and breathing out, keeping the arrangements of the method, increasing or decreasing them. When all things, in a word, come to harmony by division and union, without the methods being neglected in any

62

way, the nature is transformed. For the nature being turned upon itself is transformed; and it is the nature and the bond of the virtue of the whole world.

And that I may not write many things to you, my friend, build a temple of one stone, like ceruse in appearance, like alabaster, like marble of Proconnesus, having neither beginning nor end in its construction. Let it have within it a spring of pure water glittering like the sun. Notice on which side is the entry of the temple and, taking your sword in hand, so seek for the entry. For narrow is the place at which the temple opens. A serpent lies before the entry guarding the temple; seize him and sacrifice him. Skin him and, taking his flesh and bones, separate his parts; then reuniting the members with the bones at the entry of the temple, make of them a stepping stone, mount thereon, and enter. You will find there what you seek. For the priest, the man of copper, whom you see seated in the spring and gathering his color, do not regard him as a man of copper; for he has changed the color of his nature and become a man of silver. If you wish, after a little time you will have him as a man of gold.

LESSON 2

1. Again I wished to ascend the seven steps and to look upon the seven punishments, and, as it happened, on only one of the days did I effect an ascent. Retracing my steps I then went up many times. And then on returning I could not find the way and fell into deep discouragement, not seeing how to get out, and fell asleep.

And I saw in my sleep a little man, a barber, clad in a red robe and royal dress, standing outside the place of the punishments, and he said to me "Man, what are you doing?" And I said to him, "I stand here because, having missed every road, I find myself at a loss." And he said to me "Follow me." And I went out and followed him. And being near to the place of the punishments, I saw the little barber who was leading me cast into the place of punishment, and all his body was consumed by fire.

2. On seeing this I fled and trembled with fear, and awoke and said to myself "What is it that I have seen?" And again I reasoned, and perceiving that the little barber is the man of copper clothed in red raiment, I said "I have understood well; this is the man of copper; one must first cast him into the place of punishment."

Again my soul desired to ascend the third step also. And again I went along the road, and as I came near to the punishment again I lost my way, losing sight of the path, wandering in despair. And again in the same way I saw a white-haired old man of such whiteness as to dazzle the eyes. His name was Agathodæmon, and the white old man turned and looked on me for a full hour. And I asked of him, "Show me the right way." But he did not turn towards me, but hastened to follow the right route. And going and coming thence, he quickly gained the altar. As I went up to the altar, I saw the whitened old man and he was cast into the punishment. O gods of heavenly natures! Immediately he was embraced entirely by the flames. What a terrible story, my brother! For from the great strength of the punishment his eyes became full of blood. And I asked him, saying, "Why do you lie there?" But he opened his mouth and said to me "I am the man of lead and I am undergoing intolerable violence." And so I awoke in great fear and I sought in me the reason of this fact. I reflected and said "I clearly understand that thus one must cast out the lead, and indeed the vision is one of the combination of liquids."

WORK OF THE SAME ZOSIMUS

Lesson 3

1. And again I saw the same divine and sacred bowl-shaped altar, and I saw a priest clothed in white celebrating those fearful mysteries, and I said "Who is this?" And, answering, he said

to me "This is the priest of the Sanctuary. He wishes to put blood into the bodies, to make clear the eyes, and to raise up the dead."

And so, falling again, I fell asleep another little while, I mounted the fourth step I saw, coming from the East, one who had in his hand a sword. And I saw another behind him, bearing a round white shining object beautiful to behold, of which the name was the meridian of the Sun,[2] and as I drew near to the place of punishments, he that bore the sword told me, "Cut off his head and sacrifice his meat and his muscles by parts, to the end that his flesh may first be boiled according to method and that he may then undergo the punishment." And so, awaking again, I said, "Well do I understand that these things concern the liquids of the art of the metals." And again he that bore the sword said "You have fulfilled the seven steps beneath." And the other said at the same time as the casting out of the lead by all liquids, "The work is completed."[3]

The process here symbolized is probably the same as that which "Cleopatra" was writing, the chemical reaction between metals and a chemical reagent and the subsequent restoration to a metallic condition; Zosimus, we feel, is most deeply impressed by the destruction of the metals and the violence they undergo in chemical action, and, looking at the chemical process in this way, it becomes filled with meaning. As the artist can see in a landscape graciousness, solemnity, terror, so Zosimus sees the rigor of death and pains of purgation in the turbid seethings of the alchemical vessel.

Greek alchemy, then, at least as early as about 300 A.D. and probably from its earliest period, contains the essential

[2] Or of Cinnabar.

[3] Berthelot, *Collection des anciens alchimistes grecs*, texte grec, pp. 107-112, 115-118.

features of alchemy as we know it later: its secrecy, its symbolic character, the correspondence of the operations in the vessel to those of the greater world, the universal spirit that is the chief agent. The principal feature that is lacking is that of the elixir or philosopher's stone. The medieval alchemist was seeking a substance of enormous potency, a small quantity of which would transform a very much larger quantity of base metal into silver or gold, and which had unexampled powers of healing the human body and indeed of perfecting all things in their kind.

This idea is not developed, perhaps not present at all, in the works of the Greek alchemists, whose object generally seems to be the making of gold, not of a marvelous substance which had the power of transforming a metal into gold. There is a certain amount of discussion of "the medicine" (*pharmakon*) and also of "the stone that is not a stone," but this seems to relate to some substance useful in the Art and not to the unique and potent substance that was later called "the stone."

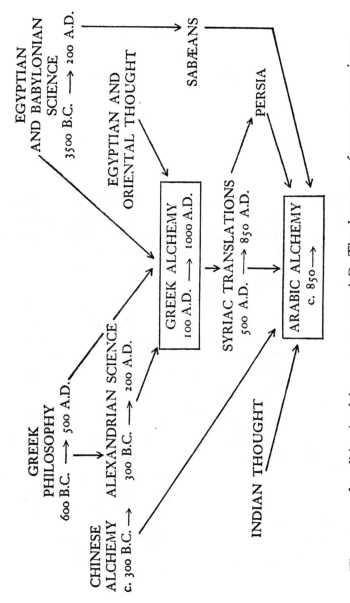

The stream of tradition in alchemy up to 1000 A.D. The dates are, of course, approximate.

VI

Chinese Alchemy[1]

IT is remarkable that, at a period some two or three centuries before the earliest Greek writings on alchemy, there appeared in China accounts of beliefs and processes which must be called alchemical. It may at once be said that the evidence is not sufficient to enable us to decide whether Chinese alchemy derived from Western alchemy or *vice versa*, or whether both arose from a single source, though none of these hypotheses about its origin is to be rejected out of hand. In our present state of knowledge we must treat this early Chinese alchemy as showing remarkable parallelism to Western alchemy but not as being connected with it by any known contacts.

As we shall see, the alchemy of China is primarily concerned with the prolonging of life. The idea of a drug which could act as an elixir of immortality is found in Indian literature before 1000 B.C., and there are some hints of the idea of alchemy in the Atharva-veda[2] which belongs to the

---

[1] This section owes much to a paper by Prof. Homer H. Dubs, "The Beginnings of Alchemy" (*Isis*, Vol. 38, Parts 111-112, p. 75) which the reader may consult.

[2] *Veda*, the most ancient sacred literature of the Hindus, comprises more than 100 extant books; the *Atharva-veda* is the fourth and apocryphal veda, mainly composed of incantations preserved in two versions.

same period. It is possible, but by no means certain, that this was the source of Chinese alchemy.

The first evidence of alchemy in China is an indication that it was practiced by Dzou Yen as early as the fourth century B.C. It is certain that a law was enacted against the practice of counterfeiting gold by alchemical methods in the year 175 B.C. and it is obvious that alchemy must have existed for no little time before it could have become a scandal requiring legislation to repress it. Despite this law it existed or at any rate was discussed at the Imperial Court about 130-120 B.C. In the year 60 B.C. the Emperor appointed a well-known scholar, Liu Hsiang, as Master of Recipes in order that he might prepare alchemical gold and so prolong the Imperial Life. He failed to make the gold and was disgraced.

These facts are indisputable evidence of the early practice of alchemy in China—much better evidence than the names of authors at the head of alchemical texts which may prove to be false attributions.

There are plenty of charming legends about the early masters of Chinese alchemy which serve at least to show the attitude of their successors towards the origin of their Art. An example is the story of Wei Po-yang, who lived in the present province of Kiangsu about 120 A.D. One of the Chinese biographical encyclopedias tells us that Wei Po-yang "entered the mountains to make efficacious medicines. With him were three disciples, two of whom he thought were lacking in complete faith. When the medicine was made, he tested them. He said, 'The gold medicine is made but it ought first to be tested on the dog. If no harm comes to the dog, we may then take it ourselves; but if the dog dies of it, we ought not to take it.' (Now Po-yang had brought a white dog along with him to the mountain. If the number of the treatments of the medicine had not been sufficient or if harmonious compounding had not reached the required standard, it would contain a little poison and would cause temporary death.)

"Po-yang fed the medicine to the dog, and the dog died an instantaneous death. Whereupon he said, 'The medicine is not yet done. The dog has died of it. Doesn't this show that the divine light has not been attained? If we take it ourselves, I am afraid we shall go the same way as the dog. What is to be done?' The disciples asked, 'Would you take it yourself, Sir?' To this Po-yang replied, 'I have abandoned the worldly route and forsaken my home to come here. I should be ashamed to return if I could not attain the *hsien* (immortal). So, to live without taking the medicine would be just the same as to die of the medicine. I must take it.' With these final words he put the medicine into his mouth and died instantly.

"On seeing this, one of the disciples said, 'Our teacher was no common person. He took the medicine and dies of it. He must have done that with especial intention.' The disciple also took the medicine and died. Then the other two disciples said to one another, 'The purpose of making medicine is to attempt at attaining longevity. Now the taking of this medicine has caused deaths. It would be better not to take the medicine and so be able to live a few decades longer.' They left the mountain together without taking the medicine, intending to get burial supplies for their teacher and their fellow disciple.

"After the departure of the two disciples, Po-yang revived. He placed some of the well-concocted medicine in the mouth of the disciple and in the mouth of the dog. In a few moments they both revived. He took the disciple, whose name was Yü, and the dog, and went the way of the immortals. By a wood-cutter whom they met, he sent a letter of thanks to the two disciples. The two disciples were filled with regret when they read the letter." [3]

It is clear that alchemy, despite all prohibitions, flourished in China exceedingly during later centuries, and it is very probable that the Arab alchemists received some information about it. It is certainly notable that the idea of the elixir as a medicine prolonging life was present to the Arabs and not to their Greek-speaking predecessors.

By the sixth century A.D. Chinese alchemy had begun to decline. We find the same phenomenon as later occurred in Europe (Ch. XIV), namely the transformation of alchemy from a practical art to a mystical exercise, and the belief that the old texts, which were certainly intended to be practical directions, were allegories concealing spiritual truths.

Chinese alchemy is far from identical with Western. In both we find the idea of transmutation and the making of gold; gold for the Chinese was not a medium of currency but an imperishable substance, and so the whole emphasis of the Chinese alchemists is upon the making of gold as a substance to confer longevity or immortality in the body, an idea which does not appear to enter Western alchemy until the Islamic period. The following verses clearly express the intentions of the Chinese alchemists:

> If even the herb *chü-sheng* can make one live longer,
> Why not try putting the Elixir into the mouth?
> Gold by nature does not rot or decay;
> Therefore it is of all things most precious.

[3] Wu and Davis, "The Ts'an T'ung Ch'i of Wei Po-yang," *Isis*, Vol. XVIII, 2, No. 53 (1932), p. 214.

The Alchemists

When the artist (i.e. alchemist) includes it in his diet
The duration of his life becomes everlasting . . .
When the golden powder enters the five entrails,
A fog is dispelled, like rain-clouds scattered by wind.
Fragrant exhalations pervade the four limbs;
The countenance beams with well-being and joy.
Hairs that were white all turn to black;
Teeth that had fallen grow in their former place.
The old dotard is again a lusty youth;
The decrepit crone is again a young girl.
He whose form is changed and has escaped the perils
    of life,
Has for his title the name of True Man.[4]

It was believed that artificial gold was a substance of such potency that the eating of food from vessels made of it would be conducive to longevity. It was further supposed that a "pill of immortality" could be prepared. We thus see at the outset a difference of intention between the Chinese and Western alchemy.

The Chinese alchemists were followers of Lao Tzū, whose profound philosophy, Taoism, given to the world in the sixth century B.C., rapidly became associated with all manner of wonder-working and magic. Its attention was early focussed on the problem of mortality. By bringing the body into a perfect harmony with the Tao, the "way of the universe," it would acquire the attributes of Tao and so become deathless. This attainment of harmony with the Tao was a mystical process, possible only for men of great spiritual gifts. We may conjecture that those who did not possess those gifts, but urgently desired to live, were ready to

[4] From the 32nd section of the Pao P'u Tzu (340 A.D.) quoted by A. Waley in "Notes on Chinese alchemy," *Bulletin of the School of Oriental Studies*. London Institution, Vol. VI, Part I, 1930, p. 11.

72

take the short cut offered by *lien tan* "the drug of transmutation."

How was this drug to be made? The Chinese theory of matter posited two principles, *Yang*, the active or male element, and *Yin*, the female and passive element. Substances rich in *Yang* were those that imparted life and caused longevity. The most highly reputed of these was cinnabar (native red mercuric sulphide); gold was next to it in potency. We may suppose that the red color of the former was connected with the red blood of health, and that its power of forming liquid mercury, the "living" metal ("quick"-silver) also entered into it. It must have gradually appeared that cinnabar did not confer immortality, and, in typical alchemical fashion, the goal was transferred to a divine or esoteric drug or elixir, or to an alchemically prepared gold. The directions for the preparation of the elixir are obscure, but we may note that it underwent the same changes in color as the philosopher's stone of the West, namely first to white and then to red.

The process of transmutation of cinnabar into gold to be used for prolonging life seems to go back to the second century before Christ. The Chinese, like almost all men of the pre-scientific period, supposed that minerals matured in the rocks gradually becoming more precious. Cinnabar was thought to change into lead, lead to silver, silver to gold. It did not seem unreasonable that this process should be realizable in the laboratory. The method of attempting transmutation differed from that of the West. The Chinese employed boilings and fusions, chiefly; they were certainly acquainted with sublimation, which they used in making vermilion. If the translations are correct, they were also acquainted with some kind of distillation. The accounts of

73

their processes are not such as to enable us to discover what they used. Thus the following, taken from the *Ts'an T'ung Ch'i* by Wei Po-yang (c. 120 A.D.), evidently describes some process in which a solution is evaporated and crystallizes.

Above, cooking and distillation take place in the caldron; below, blazes the roaring flame. Afore goes the White Tiger leading the way; following comes the Grey Dragon. The fluttering *Chu-niao* (Scarlet Bird) flies the five colors. Encountering ensnaring nets, it is helplessly and immovably pressed down and cries with pathos like a child after its mother. Willy-nilly it is put into the caldron of hot fluid to the detriment of its feathers. Before half of the time has passed, Dragons appear with rapidity and in great number. The five dazzling colors change incessantly. Turbulently boils the fluid in the *ting* (furnace). One after another they appear to form an array as irregular as a dog's teeth. Stalagmites which are like midwinter icicles, are spit out horizontally and vertically. Rocky heights of no apparent regularity make their appearance, supporting one another. When *yin* (negativeness) and *yang* (positiveness) are properly matched, tranquility prevails.[5]

The notion of the philosopher's stone, that is to say of a substance of which a small quantity will transmute a much larger quantity of base metal into gold or silver, is also first heard of in the Chinese texts. In a text dating from about the beginning of the Christian era we are told that "a gentleman of the Yellow Gate at the Han [imperial court], Cheng Wei, loved the art of the Yellow and White [alchemy]. He took a wife and secured a girl from a household which knew recipes . . . [Cheng] Wei [tried to] make gold in accordance with 'The great Treasure (*Hung-bao*)' in the

[5] Wu and Davis, *op. cit.*, p. 258.

74

pillow [of the King of Huai-nan, but] it would not come. His wife however came and watched [Cheng] Wei. [Cheng] Wei was then fanning the ashes to heat the bottle. In the bottle there was quicksilver. His wife said, 'I want to try and show you something.' She thereupon took a drug out of a bag and threw a very little into [the retort]. It was absorbed and in a short while she turned out [the contents of the retort]. It had already become silver. [Cheng] Wei was greatly astonished, and said, 'The way (*Dao*) [of alchemy] was near and was possessed by you. But why did you not tell me sooner?' His wife replied, 'In order to get it, it is necessary for one to have the (proper) fate.' " [6]

The parallel between Chinese and Western alchemy is certainly remarkable, but the fact that the former was chiefly a means of prolonging life and the latter mainly a means of obtaining wealth seems to rule out the possibility of one having been derived from the other. That the Chinese tradition contributed to Western alchemy by way of Islam the idea of an elixir of life is, however, quite probable; the necessary contacts between China and Islam certainly existed, and it would be very surprising if advantage had not been taken of them by the latter's alchemists, who will be discussed in the following chapter.

[6] Dubs, *op. cit.*, p. 78.

VII

Alchemists of Islam

EVEN while Greek science was at its zenith, there
existed other scientific cultures in the Near and
Middle East. In India and Persia, and among the
Sabæans of the eastern parts of Syria, much attention was
given to astronomy and mathematics, concerning which,
however, there remain but scanty records. The important
thing for posterity is that, during the five centuries after
the birth of Christ, in these countries a body of natural
philosophers was ready to receive and cultivate new knowledge. The most vital of these centers was Syria, which was
a true meeting-ground of cultures and tongues. Latin,
Greek, Syriac, Persian, and, after the rise of Islam, Arabic,
were all current languages, and so Greek learning was able
to take root there, to draw new life from the fertilizing mixture of cultures and to traverse the whole of the Near East.

The immediate cause was the expulsion of the learned
sect of the Nestorians from Constantinople in 431 A.D.
They formed an active school of Greek learning at Edessa
in the north of Syria. Thence they were expelled by the
Greek Emperor in 489. They then moved to Nisibis in
Mesopotamia, and soon after 500 A.D. finally settled at
Jundai-Shāpūr, the great Persian medical school, some dis-

76

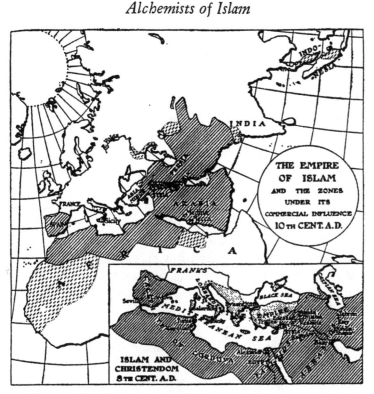

Fig. 11.—The Islamic world. (Courtesy of *The Geographical Magazine*, London, England.)

tance north of Basra. The Nestorians long retained a knowledge of Greek and soon began to translate Greek works into Syriac. In the next century the Monophysite Christians were also expelled from Constantinople and migrated to Syria and Persia. Some, at least, of the Greek works on alchemy were translated by them into Syriac.

Between the years 622 A.D. and 750 A.D. the Arab states and wandering tribes, united in the religious enthusiasm of Islam, conquered and imposed their way of life upon Asia

Minor, Syria, Persia, Egypt, Africa, and Spain. At first their attitude to the infidel learning was hostile; but after 750 A.D., under the Abbasid caliphs of Baghdad, they developed a thirst for learning. Thereafter Greek works of philosophy, mathematics, and science could hardly be translated quickly enough to satisfy them. We do not yet know very much about the attitude of Islam to alchemy in the earlier period, though certainly much is to be discovered when more texts are examined, but we know that there was great activity therein soon after 900 A.D.

The great figure among the alchemists of Islam—indeed one of the few alchemists of whom the educated man has heard—is Geber. European writers have looked on him as the founder of the Art; but recent researches seem to show that the attribution of alchemical books to him is another case of attribution of books from many hands to a single famous legendary figure.[1]

It is a curious story. The Imāms were the spiritual and secular heads of Islam. The sixth Imām was Ja'far b. Muhammad al-Sādiq, who was exalted by the Shi'ites into the position of the great possessor of the secret sciences, notably alchemy and astrology. Numerous works are attributed to him, but in fact are forgeries of later date. He was supposed to have had a pupil, Abū Mūsā Jābir b. Hayyān, who flourished about 760 A.D. Under this name there appear very numerous treatises. Most of these are alchemical, but others treat of medicine, astronomy, astrology, magic,

---

[1] It should be said that the views of the late Paul Kraus, indicated in this section, are not accepted by some experts, such as H. E. Stapleton, who regard Jābir as a real person and practicing alchemist, and suppose that the works that bear his name were written by him, though extensively re-edited in the ninth century.

mathematics, music, philosophy, and indeed constitute an encyclopedia of the sciences. This Jābir is the figure who appears in the medieval writings (and in the works of the earlier historians of chemistry) as "Geber, King of the Arabs." It has recently been maintained that no Arab author mentions Jābir till two centuries after the time at which he was supposed to have lived. It is now regarded as very probable that this vast body of writings was composed by members of a group resembling in its religious leanings the secret sect of natural philosophers who called themselves Ikwān al-safā, which has been variously translated "Brethren of Purity" or "Faithful Friends."

The Brethren of Purity composed an encyclopedic collection of letters much resembling the Jābirian writings. We may suppose, then, a sect with a strong belief in the power of science to purify the soul, who ascribed the works of its members to the legendary Jābir—which is somewhat as if an encyclopedia of the sciences were to be written today by some Communist secret society and published under the name of Voltaire. The practice of attributing the books of a school of workers to the hand of the master was, however, common in antiquity. The authors of the Jābirian treatises were deeply impressed with the possibilities of science. It is true that they included in science much that we should call magical; but we must credit them with a belief in the power, not merely of book-learning, but of practical laboratory operations. It is true that their theories and practice went far astray from truth and usefulness, though they discovered some useful things without realizing their value. Indeed, all through its history, alchemy continued to make useful physical discoveries in the attempt to perform the physically impossible.

Jābir (for we will keep the name to denote the authors of the writings attributed to him) was certainly acquainted with the work of the Greek alchemists, though doubtless he read them only in translations or paraphrases. The writings of the Greek alchemists as we have them today are a mass of fragments, and we are sure that the Arabs could have read much that is not available to us. We, therefore, cannot say how much of Jābir's work is original, but we can say that we find in it much that is not contained in the Greek alchemy that we know.

As is usual in the works of the alchemists, we see a consistent theory of the manner in which gold should be made, and a process based on that theory but incapable of giving rise to the effects intended. Jābir divided the substances he knew into:

1) *Spirits*: volatile bodies such as camphor, sal ammoniac, mercury, arsenic, and sulphur.
2) *Metallic* bodies: the metals.
3) *Bodies*: non-volatile pulverizable solids, i.e., substances other than "spirits" or metallic bodies.

This classification comes from the Greeks who regarded the metals as combinations of a body and a soul or spirit. There were however other systems of classification in which mercury was classified as a metal.

Jābir's theory of the formation of the metals is clearly derived from Aristotle (pp. 12-13), whose "moist" and "dry" vapors have now become vapors of mercury and sulphur. These combine in the rocks and produce the metals; the different metals differ only according to accidental qualities, the cause of the difference in which is the varying quality of the "sulphur." The Greek alchemist spoke of several different sulphurs, and the word, before the period of modern chemistry, was a wide term for a fusible, volatile, combustible body. However, sulphur in the modern sense was also very well known; but there were thought to be many varieties of it: yellow sulphur, white sulphur, green sulphur, black sulphur, etc., which may have been specimens of the element discolored by various impurities, but may also have been chemical compounds containing, or apparently containing, sulphur. Miners in some localities speak of iron pyrites as "sulphur," and Jābir's sulphurs need not be taken to be any more like sulphur than is that mineral.

This notion that the metals are composed of mercury and sulphur remained a part of alchemy and chemistry even into the eighteenth century. The idea of the presence of an "inflammable principle"—sulphur—in metals and indeed in almost all bodies is the progenitor of the notion of phlogiston.[2] Chapter XIII gives some account of the meaning which the words "sulphur" and "mercury" had assumed by the eighteenth century. Though Jābir thought metals to be made from mercury and sulphur, he also supposed them to be composed ultimately of the four elements, earth, water, air, and fire, and to have the qualities of these elements—dryness, cold, moisture, and heat—in varying proportions. A metal was supposed to have one pair of qualities externally and another pair internally. Thus we have:

	Outer Qualities	*Inner Qualities*
GOLD	Hot—moist	Cold—dry
SILVER	Cold—dry	Hot—moist

To turn silver to gold it was necessary to turn its nature inside-out, so to speak. In order to bring about a transmuta-

[2] Cf. p. 211.

tion, the alchemist, according to Jābir, had to alter the proportions of these qualities of heat, moisture, cold, and dryness. This did not seem at all unreasonable or difficult, for the idea of altering them was already very common in medicine.

The Greeks, and especially Galen, whose works were very well known to the Arabs, attributed many diseases to excess of one of these qualities. If a patient suffered from an excess of, e.g., the hot element, he was given a medicine made of substances in which the cold elements were supposed to preponderate. Jābir tried to do the same—to cure the baseness of metals with medicines, which he termed *elixirs*. The Greek alchemists likewise talked of the "medicine" (*pharmakon*), which was to be added to a mixture in order to cause a transmutation. Jābir greatly developed the idea of the "supreme elixir," the medicine of the metals, and invented his method of the balance—a very systematic way of tackling the problem of transmutation.

First, he considered that the alchemist should be able to find out the proportions of earth, water, air, and fire in any body, and then adjust these proportions, so as to convert it into another body, by adding an elixir made of the pure elements combined in the right porportion to supplement and correct the metal's deficiencies or excesses. This quantitative idea sounds most scientific and modern, but actually it was not founded on laboratory measurement, for metals could not be analyzed into anything like earth, air, fire, and water, and so the proportions of these supposed elements could not be measured.

On the other hand, organic bodies could be analyzed by distillation—probably also a Greek idea. Just as Zosimus distilled eggs, so Jābir distilled all kinds of animal and vegetable products. It was in this way, probably, that sal ammoniac came to be separated from the dried dung of animals. By distilling any of these organic bodies he obtained in every case (1) a liquid, which was the element of water (cold and moist), (2) something he called oil or grease, which was an inflammable body which he identifies with the element of air (hot and moist): this was presumably a mixture of volatile combustible organic liquids and gases, (3) a combustible colored substance called fire or tincture (probably a tarry body) which he identifies with fire (hot and dry), (4) a dry mineral residue, mainly charcoal, which he identifies with the element of earth (cold and dry).

Now, each of these supposed elements has two qualities, and Jābir's aim is to make "pure elements" with only one quality, so that he may be able, for example, to add coldness to a metal without, at the same time, adding moisture or dryness. Thus he does not want common water which is cold and moist, but a water which is cold but no longer moist. To obtain this, he distils common water and redistils it repeatedly, adding substances thought to be very dry and so capable of removing the moist quality from the water. After hundreds of redistillations, the water, he tells us, becomes white and brilliant and solidifies like salt. This, he says, is a pure element, and is simply the quality of coldness residing in first matter. Similar processes applied to the other distilled products were supposed to give the warm, the moist, and the dry elements. Some of the processes he described involve as many as 700 distillations.

What are we to make of this? We have never redistilled water 700 times with the addition of the chemicals Jābir

83

mentions, so we cannot disprove his assertion; but no scientist could envisage any possibility of altering water (except in isotopic constitution) in this way. Was Jābir lying? Was he describing what he thought ought to happen, though he himself had never tried the experiment? Or did he mean something entirely different and as much or little allied to chemistry as freemasonry is to building? These questions cannot be decisively answered, though I believe that the second explanation is the most probable.

Why, we may ask, did not the alchemists who followed him discover the ineffectiveness of these processes and expose them as frauds? Largely, I think, because hardly any alchemist reached his seven hundredth distillation. Glass was not of modern standard, and I am inclined to think that long before the seventieth, let alone the seven hundredth, distillation, some accident occurred; a still cracked or a furnace collapsed or some other incident brought the work to a premature close.

Jābir supposed, then, that the alchemist could make the element that is wholly cold from his "water," the element that is wholly moist from his "oil," the element that is wholly dry from his "earth," and the element that is wholly hot from his "tincture." The last seems to have been the precursor of the philosopher's stone, being described a transparent body, brilliant, lustrous, and red. It was presumably that which was lacking in base metals and present in gold.

Having obtained these "pure elements," the alchemist was required to mix them in specified numerical proportions, so forming a suitable "elixir," which was to be applied to the metal by a somewhat complicated process. Transmutation should then take place.

Jābir's number system seems to us very odd. The Greeks had denoted variations, e.g., of hotness and coldness, moisture and dryness, by "degrees," though they had no means of measuring them quantitatively. Thus poppies were a drug "cold in the fourth degree." Jābir applies this more elaborately, assigning a "value" to each substance. Thus, e.g., if gold is worth 1, then elixir is worth 5. The power of each treatment is denoted by a special fraction. Thus a sublimation is worth $\frac{1}{50}$ and a fusion $\frac{1}{200}$. Jābir then works out equations on this basis, e.g.,

$$\text{(Gold) } 1 \times \text{(Fusion) } \frac{1}{200} \times 1000 = \text{(Elixir) } 5$$

The conclusion reached is that 1000 fusions should convert gold into elixir. Well, we can't give this reasoning any high praise, but we can see its importance in chemical history. It was called the method of the balance (mīzān) and it vigorously emphasized the importance of quantitative considerations and must have involved careful weighings-out of quantities. It may be thought to have done its part in introducing the quantitative idea into chemistry.

The authors who wrote under the name of Jābir were by no means the only important Arab alchemists. There were certainly such alchemists before the time of the Jābirian writings. Thus there is a *Book of Crates* which may date from 800 A.D. or a little later, and is little more than an Arabic adaptation of Greek alchemy. Many Arabic works on alchemy exist, but few have been translated. Some are full of mystical and secret meanings and cannot be discussed here, though we shall have more to say of similar works in Chapter XI. But one man stands out as a hard-headed and practical man of science—namely, the man who was known to the Latin world as Rhases.

The Alchemists

The name Rhases is the latinized form of Abū Bakr Muḥammed b. Zakariyā al-Rāzī, the last name being that by which he is usually known, and denoting the fact that he came from the town of Raiy in Persia. He was a Persian, and it is notable that most of the famous scholars and scientists of Islam were not of Arab origin, many of the earliest, indeed, being Nestorian Christians.

Al-Rāzī was the first of the great encyclopedists of Islam, men learned in almost every branch of science and philosophy. He wrote on all manner of medical and surgical subjects, on philosophy, alchemy. mathematics, logic, ethics, metaphysics, religion, grammar, music, chess, and draughts! By profession, he was a physician, and his medical writings were more important than his alchemical works. His life gives us a picture of a scientific man of the great days of the science of Islam.

He was born in the year 864 A.D., a period when European learning was at its lowest and Arabic learning in its golden youth. His first studies were in philosophy, and, in his early years, he was a writer of poetry. Like many men of science—Galileo is another instance—he was a skilled musical performer and even composed an encyclopedia of music. At the age of about thirty he traveled to Baghdad, where a famous hospital was located. He seems to have turned to medicine chiefly through that intellectual curiosity which was his life's passion, and he found it of such interest that he decided to devote his life thereto.

His interest in alchemy seems to have dated from his early years and he is reported to have said that "No man deserves the name of 'philosopher' unless he be a master of theoretical and applied chemistry." He appears to have gone blind from

86

cataract near the end of his life, and to have died when he
was about sixty, at a date near 923 A.D.

The only quatrain of his poetry that survives has a
poignant ring:

> This feeble form decaying day by day
> Warns me that I must shortly pass away.
> Alas! I know not whither wends the soul
> When it deserts this worn and wasted clay.[3]

Unfortunately none of his alchemical works has been
translated directly from Arabic and printed, but the sum-
maries and descriptions of them which have been made by
those who have been able to study the MSS show that he
was an extremely practical and sensible chemist. He cites
the names of numerous Greek alchemists and probably had
an acquaintance with their works at first or second hand.
His *Book of Secrets* divides chemical substances into well-
marked classes: "spirits," metallic bodies, stones, vitriols,
boraxes, and salts. He describes the equipment needed for
the art, including apparatus for distillation and sublimation,
furnaces, etc. He goes on to discuss a number of chemical
operations: the preparation of various "poisonous" waters,
which apparently included ammonia and some of the strong
acids; he describes calcinations, sublimations, dissolutions,
combustions, and finally, though in obscure terms, the making of elixirs and of gold and silver. His work cannot always
be clearly understood; but it is very free from the deliberate
concealments, the allegory and rhetoric of so many texts.
It is, in fact, the work of a man of science, dealing with a
subject of which the theory was not clearly understood.

[3] Translation by G. S. A. Ranking in "Life & Works of Rhazes," *XVIIth
International Congress of Medicine*, 1913, Sec. XXIII, p. 237.

The Alchemists

One more Arab chemist may engage our attention, namely, Abu'l-Qāsim al'Irāqī, who, probably in the thirteenth century, wrote a work called *Knowledge Acquired Concerning the Cultivation of Gold*. The English translation by E. J Holmyard gives an opportunity for anyone who is so fortunate as to possess the somewhat rare book to see for himself what Arabic alchemy is like.

Al'Irāqī's theory of alchemy is much like Jābir's. He adopts Aristotle's theory of the genesis of metals and supposes that the base metals are imperfect varieties of gold, and that their properties can be modified by the red or white elixir so as to make them true gold or silver.

He instances as a proof of the possibility of such change the true fact that lead, heated for a long time in the fire, yields a small amount of silver, which, as we know, though the Arabs could not, is derived from the small proportion of silver compounds present in all lead ore, but which al'Irāqī supposed to have been produced by transmutation.

His theory is excellent, but when he comes to tell his readers what to do, he expresses himself in dark sayings and allegories, and quotations from the various alchemical sages. We have the impression, indeed, that he is a man with a clear grasp of a theory concerning alchemy, but without successful experience of the practice.

The Arabic alchemists passed on to the Western world not only their chemical knowledge and practical technique, but also a great deal that has no relation to chemistry as we know it. One of the most famous of these is the brief writing called the *Emerald Table of Hermes*. Hermes is a name often met with in late Greek literature, and perhaps the most ancient alchemical writings are the few fragments to which

88

that name is attached. Further fragments are preserved in, Arabic writings, and some of these clearly derive from a Greek original. The *Emerald Table* may be one of these though no Greek version has as yet been found. It had an enormous influence on the later alchemists and is therefore worth repeating in full. The key to this strange document is the doctrine of *pneuma* and if the reader looks at pages 11-16 of this book, he will see what the author intends. Here is one of the several versions of the *Table*.

THE WORDS OF THE SECRET THINGS
OF HERMES TRISMEGISTUS

1. True, without deceit, certain and most true.
2. What is below, is like what is above, and what is above is like that which is below, for the performing of the marvels of the one thing.
3. And as all things were from one thing, by the mediation of one thing: so all things were born of this one thing, by adaptation.
4. Its father is the Sun, its mother is the Moon; the wind carried it in its belly; its nurse is the Earth.
5. This is the father of all the perfection of the whole world.
6. Its power is integral, if it be turned into earth.
7. You shall separate the earth from the fire, the subtle from the gross, smoothly and with great cleverness.
8. It ascends from the earth into the heaven, and again descends into the earth and receives the power of the superiors and inferiors. So thus you will have the glory of the whole world. So shall all obscurity flee from thee.
9. This is the strong fortitude of all fortitude: because it will overcome every subtle thing and penetrate every solid.
10. Thus was the earth created.
11. Hence will there be marvelous adaptations, of which this is the means.

12. And so I am called Hermes Trismegistus, having three parts of the Philosophy of the whole world.
13. What I have said concerning the operation of the Sun is finished.

This enigmatic work evidently conveyed to the reader that the operation of the Sun (the sign for Gold) was carried out by a "spirit," universal, the source of all things, having the power of perfecting them. Its virtue is integral (i.e., having the power to convert the diverse into a single substance), if it be turned into earth (i.e., solidified). This conveyed that the "stone" was to be a solidified *pneuma*. *Pneuma* was the link between earth and heaven, having the virtue of the celestial and subterranean regions—the power of the whole cosmos from the fixed stars to the center of the earth. It overcomes every nature and penetrates every solid. It is the source of the whole world and so it can be the means of changing things in a wonderful way. The three parts of the philosophy of the whole world are presumably of the celestial, terrestrial, and subterranean regions.

This document was translated into Latin before 1200, and it is one of the most important sources of medieval alchemy. The ideas it conveyed fitted in very well with the science of the time and although it gave no clear idea as to how the *pneuma* was to be fixed as a solid, yet it set this up as a goal. Much of medieval alchemy consisted in fact of variations on this theme.

It is not easy to sum up what the Moslem world did for alchemy and chemistry. We do not know how much it received from the Greek alchemists, but it was certainly much more than survives in the Greek writings which have come down to us. We do not know how much the Arabs discovered, because so many of their works are still unstudied, and in no field of the history of science is research more urgently needed. We do not even know how much they handed on to the Western world, because we are not sure of the genuineness of many texts that purport to be translations from the Arabic.

But it seems clear that by the twelfth century the Arabs, in addition to what the Greeks had taught them, knew the preparation of sal ammoniac, ammonia, the mineral acids and borax. Consequently, the total chemical knowledge of the Moslem world in the twelfth century was quite considerable. The methods of distillation and other chemical operations; a considerable number of important chemical preparations; the idea of transmutation by a medicine or a "stone"; the lore of the four elements—all this and much more awaited revelation to the Latin-speaking Western world, which, as far as we can see, knew nothing of alchemy and whose pharmacy and metallurgy consisted only of the simplest poundings, strainings, boilings, and meltings.

One of the most important means of transmitting this knowledge was the collection of works attributed to "Geber, the most famous Arabian Prince and Philosopher." We have seen that the Arabic texts attributed to Jābir are considered to be a compilation made many years after the time when he is supposed to have lived. But these Arabic treatises are not identical with the Latin works which bear the name of Geber. Of these the most important is the *Summa perfectionis*, which was the most important source for medieval alchemy and chemistry. This book is certainly derived from Arabic sources, but it does not seem to be older than the latter part of the thirteenth century. We do not know whether it is a translation of an Arabic text or a summary of Arabic chemistry compiled in Latin by a Western writer. At all events, since these Latin works of "Geber" derived their contents from the Arabs, they may be considered here.

The writer of these treatises was an alchemist, that is to say, he believed in and argued for the possibility of the transmutation of metals and indicated methods for carrying out the work. The most important features of these texts are their advocacy of the sulphur-mercury theory of metals, their description of chemical methods, and the beginning of analysis—namely the setting out of numerous methods for testing a metal to discover whether it is genuine gold. Geber's work transmitted the knowledge of the mineral acids to the Western world, and this is perhaps the most important single piece of chemical information that these books contain.

Take vitriol of Cyprus (copper sulphate, probably containing ferrous sulphate) one pound, saltpetre two pounds, and alum of Yemen (aluminium sulphate) one fourth part; extract the water, heating the alembic to redness . . . It is made much sharper if you dissolve in it a fourth part of sal ammoniac; because that dissolves gold, sulphur, and silver.

The recipe is quite sound chemically and produces nitric acid, which continued to be made by Geber's method for four centuries after his text was written. By dissolving sal ammoniac (ammonium chloride) in the acid a certain amount of chlorine is set free; the resulting acid would attack gold, which nitric acid alone does not, and would attack sulphur and silver more rapidly.

This was a discovery of prime importance, but to the alchemists and chemists of the Middle Ages, the descriptions

92

and illustrations of furnaces were probably of even greater
value. The illustration given in Figure 12, is taken from a
seventeenth-century printed edition of Geber, but probably

FIG. 12.—Distillation, as illustrated in Russell's translation of
The Works of Geber (1678).

adheres fairly closely to the drawings of some early manuscript.

The works of Geber were the medieval alchemist's textbook and vade-mecum. They are very clear and free from
mystery—but they did not enable their readers to make

gold. There was, indeed, something of a reaction against them. The philosophic and mystical type of alchemist looked down on the busy laboratory workers intent on making gold from ordinary materials by chemical methods and dubbed them "Geber's cooks." Some, indeed, affected to find in Geber an author who knew the secret and hid it under a mass of practical instructions which, literally interpreted, led nowhere.

Although the works of Geber are very far from being the first alchemical texts available to Western Europe, and although the majority of them have no known Arabic original, yet they can be truly regarded as the most important means by which the Arabic chemical knowledge became available to the alchemists of medieval Christendom.

VIII

The Alchemists in Europe

WE pass to the twelfth century. Europe was becoming rapidly more organized and settled. The towns were becoming important centers; increasing trade was bringing new wealth and leisure. In the intellectual world it was the dawn of a new age. The West, whose learning and intellectual activity had been almost entirely theological, was beginning to put forth philosophers. Such men as Berengarius, Roscelin, Anselm, Abelard, Hugh of St. Victor were exercising themselves on philosophical problems, chiefly through the medium of the Platonic tradition that had come down in the West. Universities were beginning to take shape; there was an appetite for learning and a thirst for new material. That material was in the hands of the Arabs and of the Jews who dwelt in the Moslem world.

What the Western world required of Islam was its knowledge of philosophy and science, and it is remarkable how large a proportion of the first books to be translated were scientific. Western Europe knew scarcely anything of medicine, its astronomy and mathematics were rudimentary, chemistry and physics scarcely existed for it. The only works concerned with anything like alchemy were certain books of technical recipes, such as the *Compounds fo-*

95

Colouring, *The Key to Painting*, and *Book of Fires*, which went right back through the Byzantine tradition to the recipes of the papyri of Leyden and Stockholm. They gave information about all manner of practical matters, such as dyeing, the making of pigments, treatment of metals, and so forth. They are not strictly alchemical but are concerned with matters that interested the early alchemists.

It was obvious to those who came in contact with the frontiers of the Latin and Arab worlds in Sicily, Southern Italy, and Spain, that the Arabs were enormously superior to the "Franks" in all knowledge and skill concerning natural things, and the obvious step was to procure translations of their books. The men of the West knew neither Greek nor Arabic, the Arabs generally did not know Latin; but there were many learned Jews who knew some tongue that the "Franks" understood. By their aid, translations came to be made. There was a medical school at Salerno in Southern Italy where translations were prepared or, at any rate, used as early as the eleventh century, but the first known version of an alchemical work was made by one Robert of Chester in 1144. By 1200 some half-dozen texts had been translated, including the *Book on Alums* ascribed to al-Rāzī, and the *Emerald Table*. Interest in the subject began to grow, and in the thirteenth century alchemy was seriously discussed and also widely practiced.

The best minds of the time doubted whether alchemy was a true science or a fraud. St. Albert (Albertus Magnus), Roger Bacon, and St. Thomas Aquinas all discuss the question. Albertus, who wrote an excellent work on minerals and had gone to the mining districts to see the processes for himself, performed the practical experiment of testing some alleged alchemical gold, but found that five or six treatments in the furnace reduced it to ashes. A similar story, by the way, is told of al-Rāzī, that some of his alchemical gold tarnished after many years, and that he took it back again. Doubtless many of these so-called golds were brassy alloys like the modern ormolu and pinchbeck, or possibly mixtures of these with some gold.

Albertus Magnus was inclined, therefore, to think that the alchemists with whom he was acquainted did not make true gold. Avicenna also (980-1036) said very much the same thing: that the alchemists produced imitations of precious metals and did not bring about a true change. But Albertus took the claims of alchemy very seriously, for there were strong arguments in their favor. He thought, indeed, that the transmutation of metals was possible but very difficult, for if, as he believed, nature could transform sulphur and mercury into metals by the aid of the sun and stars, it seemed reasonable that the alchemist should be able to do the same in his vessel.

About the same time (c. 1250) Roger Bacon, who was quite certainly a practical laboratory worker and a great exponent of the merits of science, discussed the same subject. He likewise believed in the Art. In his *Opus Tertium*[1] he distinguishes speculative alchemy, a knowledge of the properties of bodies and their generation and changes—very nearly what we call chemistry—from the operative or practical alchemy which teaches man

how to make noble metals, and colors, and many other things better and more copiously by art than by nature. And this science is more important than all that have preceded it because

[1] Roger Bacon, *Opera quaedam hactenus inedita,* ed. J. S. Brewer. Rolls series. London, 1859, Vol. I, Ch. XII, pp. 39-40. The subject is also discussed in most of his other works.

it is productive of more advantage. It not only provides money and infinite other things for the State, but teaches the discovery of such things as how to prolong human life as far as nature allows it to be prolonged.

He further says that many who attempt alchemy do so without a proper knowledge of technique—distillation, sublimation, calcination, separation—but for all this, I do not think that we can be certain that he gave much attention to this art.

Every famous man who mentioned anything like alchemy, had fathered upon him alchemical works which he did not write. Thus to Aristotle, Avicenna, Albertus Magnus, and St. Thomas Aquinas were attributed treatises of which no one can believe them to be the authors. For this reason we should want some strong evidence to persuade us that the several alchemical treatises attributed to Roger Bacon were actually written by him. Moreover, his undoubted works show that he was not slow to talk of his laboratory work and observations. Alchemy, when he was writing, was not forbidden to friars. If he had engaged in the practice of the Art, he would almost certainly have mentioned it, as he mentions his optics, but, in fact, he only records that there was such a science, extols its dignity and utility, and records a few pages of alchemical recipes.

Probably the greatest mind of that great century was that of St. Thomas Aquinas, who mentions alchemy only incidentally, but gives us very interesting information which is a clue to the medieval view. The fundamental scientific theories of the time were those of Aristotle, newly learned from translations of Arabic versions of his works, and so the passage translated on pages 12-13 seemed to the men of the

98

thirteenth century to be very important. St. Thomas, writing soon after 1250 A.D., tells us:

The chief function of the alchemist is to transmute metals, that is to say, the imperfect ones, in a true manner and not fraudulently.

In his commentary on the third book of Aristotle's *Meteorologica* (Lectio IX ad finem), he has a long discussion on Aristotle's view as to the generation of metals under the earth from a dry or smoky vapor and a moist vapor (or exhalation). He accepts this point of view but adds to it something that is not found in Aristotle, namely, that this mixture or combination requires a celestial virtue which gives the product its occult operations. The celestial virtue is the active principle, the instrumental principle is heat. Thus the metals are water, in one way, for the moist vapor might have become water if it had not been made into metal. He knows that metals can be calcined by fire into earths, but gold is so pure that there is nothing for the fire to take hold of. He then says:

The remote material of such metallic bodies is the vapor included in the stony parts of the earth, but the immediate (*propinqua*) materials of metals are sulphur and mercury, as the alchemists say: so, as in the aforesaid stony places of the earth by the mineral virtue mercury and sulphur are first generated, and then from these are generated metals according to their various mixture (*commistio*). And so the alchemists, through the true art of alchemy, (but yet a difficult art, because of the occult operations of the celestial virtue, namely the mineral virtue, which because they are hidden, are imitated by us only with difficulty)—these alchemists, by the above principles or by principles laid down by themselves, sometimes make a true generation of metals, sometimes indeed from the afore-

99

said sulphur and mercury without the generation of the exhalation: but sometimes by making the aforesaid vaporous exhalation exude from certain bodies, by the application of a proportionate heat which is a natural agent.

Thus an alchemist who took the view set out by St. Thomas would try to obtain this "vaporous exhalation" by distillation and would try to get the "celestial virtue" to work upon it. Notice how near this is to the idea of the *Emerald Table*. Everything is to be made through the *one thing*. The vaporous exhalation is a "spirit" and so is the celestial virtue that works upon it.

It is not surprising that, when the greatest savants agreed that alchemy was possible, a great number of men attempted to accomplish a work which was so interesting, so noble, and so profitable. Alchemy swept like a fever over thirteenth-century Europe, and it remained for at least three centuries the chief preoccupation of those inclined to the discovery of nature's secrets.

There was a bewildering variety of alchemical practice, but before we begin to discuss what the alchemists did and how they lived, we may try to draw a picture of the medieval alchemist.

The men of the Middle Ages did not spare their satire. Thus, some of their writings concerning their priests and friars draw a picture of men of a truly Christlike holiness, while others depict them as creatures of the most despicable trickery and hypocrisy. Likewise, from their descriptions of the alchemists we can infer on the one hand the existence of a few quiet philosophical seekers into the secrets of matter, and on the other of a great many petty swindlers duping the gullible rich with faked demonstrations. Both pictures are doubtless true.

We rarely find anything but a very brief biography of an alchemist, and many of these were written considerably after their deaths. Yet by putting two and two together, we can form some picture of the best kind of alchemist.

The medieval alchemist was in most cases a cleric, "a learned clerk," not so much because there was any connection between the religious life and alchemy, but because the great majority of those who could read and write with ease were clerics, and alchemy necessarily involved the study of books. So we may picture our alchemist as a monk, and in many cases a canon, the apparent addiction of canons to alchemy being connected with the fact that their ecclesiastical duties were light and their means considerable. He would have had a thorough education, which meant that he read, wrote, and spoke Latin, the tongue in which all nations wrote their learned works. His further education would have included the elements of the science of the time, and this would comprise Aristotle's ideas about generation and corruption.

He might have taken to alchemy for various reasons. A man of scientific tastes in the Middle Ages had the choice of being a medical man, an astronomer, or an alchemist. So the type of mind that today finds chemistry an exciting and satisfying pastime was doubtless then drawn to alchemy. Others, again, were attracted by the marvelous prospect of being able to make gold, perhaps for their own enrichment, perhaps to finance a crusade or build hospitals and churches. Such works were often done by private people, not only from pure charity but because they brought to their doers great renown and the hope of prayers from those they had

101

benefited, which should release them the sooner from purgatory. To some, again, alchemy appealed as a noble work, the perfection of nature, for, as Norton tells us in his *Ordinall*,[2]

> Also it is a worke and Cure divine
> Foule Copper to make Gold or Silver fine.

Our would-be alchemist might well have found a preceptor at work in the monastery, for there were so many and oft repeated prohibitions of the practice of alchemy by monks that we can be sure it was a very common pursuit in monasteries. There was not thought to be anything wrong about honestly-conducted alchemy, but it was not always in good odor because of the many swindlers who occupied themselves with it. Of them we shall have more to say. Furthermore, the function of a friar or monk was to promote the ends of religion, and alchemy was felt to be at best a somewhat indirect way of so doing. But, whether or no there was an alchemist in his own community, he would almost certainly find or hear of someone who was practicing the Art and who might, or might not, know much about it.

He would not be able to go out and buy the alchemical books, because the art of printing had not yet made such things easily accessible. But he would probably copy for himself or have copied one or two alchemical manuscripts which someone would lend him. It seems probable that many medieval alchemists were acquainted with only a few of the writings of those who went before them.

We usually hear that at the outset the beginner was put into complete confusion by the obscurity of the alchemical

[2] In Elias Ashmole, *Theatrum chemicum Britannicum*. London, 1652.

books, and that he spent a great deal of time and money on false trials. Indeed, it is probable that by far the greatest number of alchemists gave up the search, very much the worse in pocket.

How did he set up his laboratory? There were many difficulties. First of all, it was not a good thing to be known to be an alchemist. Your neighbors, who were commonly unlearned men, took you for a wizard, a conjurer. Your ecclesiastical superior might think you were occupying your time unprofitably. If you were an alchemist who was reputed to have made gold you were in danger of being imprisoned by a local potentate or even your sovereign and set to make gold—which was not a thing that anyone claimed to be able to do to order. So alchemy was a somewhat hidden pursuit in the Middle Ages.

It was also an expensive one, for we are always hearing of people who have expended their fortunes upon it. There were no "laboratory furnishers" then, but the glassmakers and potters, who made the instruments for apothecaries and physicians, could supply the stills and glasses required, some of which called for some fairly skilled glass-blowing. There was always a trade in drugs and pigments, and those who could pay could obtain mercury, sulphur, yellow arsenic, alum, vitriol, borax, to say nothing of vinegar, wine, honey, oil, scales of iron,[3] and the various substances used in the different trades. The quality of apparatus and drugs must be supposed to have been low. No medieval glass apparatus has survived, but none of it is likely to have been better than that of the eighteenth century, which was thick and easily broken by heat. The loss from breakage was probably very high.

[3] Ferroso–ferric oxide.

The Alchemists

The first attempts to follow the recipes contained in the manuscripts seem to have been almost invariably unsuccessful, and in the career of every alchemist who claimed to have made the stone, there was what may be called an initiation into the secret of alchemy. At some stage, he would meet an older alchemist who, when he knew him to be a man worthy of the secret, told him something that enabled him to understand the alchemical books and to begin the long work of the preparation of the stone with some confidence of success. The alleged reason for this secrecy was the danger of entrusting a secret of such potentialities to any man not proved worthy of it. For the philosopher's stone meant the possession of unlimited wealth, the healing of diseases, and the power of prolonging life indefinitely.

Obviously no good man would entrust this secret to anyone not well known to him, and further security was ensured by the alchemist's oath. The true alchemist never sold the secret for money and it was only imparted by him to a disciple after the latter had sworn that he would not reveal it except to one man only whom he knew to be worthy of it and who was set on the gaining, not of riches, but of knowledge.

What was this secret? Of this we know nothing. That the alchemical works were unintelligible without it no one will doubt who has read them; but what there was that, when told by one alchemist to another, could render them intelligible, we cannot guess. Yet that something was imparted in this way is quite certain. Some alchemists, such as Charnock, seem to have been instructed in an hour or so, others much longer. Thus Thomas Norton dwelt with his master for forty days, learning the Art. As late as 1653, Elias

Ashmole, the great English antiquarian (1617-1692), records in his diary with rejoicing that William Backhouse, "lying sick in Fleet-Street, over against St. Dunstan's Church, and not knowing whether he should live or die, about eleven o'clock told me in syllables the true matter of the Philosopher's Stone, which he bequeathed to me as a legacy."

Even when the secret was gained, the work was slow, tedious, and difficult. What we know of the actual processes will be reserved for Chapter IX. It may be said here that the process was in two parts, the gross and the subtle work. The former part was a considerable and very tedious process by which the materials were gathered and purified, while the second, which was much shorter and easier, converted the materials into the stone. The process, whatever it was, needed a continuous heat which was not allowed to cease and had to be varied from a gentle warmth to something near a red heat. The alchemist had to watch his apparatus continually for months or even years together, or else entrust some part of the work to servants, who almost invariably either overheated the fire or let it go out.

There is an interesting poem by Sir George Ripley (1471) "wherein the Author declareth his Erronious experiments." He began with the obvious materials that the books appeared to dictate—sulphur and mercury, which combined to the red pigment vermilion, but came no nearer to the stone. He made "solutions" of spirits (i.e., distilled liquids), ferments, salts, iron, and steel; he worked with "waters corrosive" (i.e., mineral acids) and "waters ardent" (i.e., combustible liquids such as alcohol). He worked with sulphur, green vitriol (ferrous sulphate), arsenic and orpiment (arsenic trisulphide), and many kinds of salts, such as

sal ammoniac, potassium carbonate, borax, tartar, common
salt, saltpeter, soda. He tried working on urine, eggs, hair,
and blood, after the fashion of Jābir, and in fact ran through
all the chemicals then available:

> Of which gey tinctures I made to shew
> Both red and whyte which were untrew.

Obviously he must have observed a great variety of chemical reactions in carrying out the preparation of numerous
colored substances, distilled products, and also crystalline
salts (for he tells us

> Of Mercury and Metalls I made Chrystall Stones)

Yet he was interested only in one kind of knowledge, how
to gain the standard ends of alchemy, the white and red
stone. All the time in which he "rostyd and boyled as one of
Geber's Cooks" was lost, for none of his results were recorded. This is the measure of the distance between alchemy
and chemistry.

So much for the alchemist against whom no fraud is alleged, the man who spent a lifetime on the search for these
wonderful materials. It is very difficult to suppose such men
either to have been in bad faith or to have been deceived as
to what they did. It is still harder to believe that they succeeded, yet apparently sincere alchemical writers do claim
success and describe in detailed and pretty consistent language how they obtained the white and the red stone and
carried out transmutations, of which there remains some
written evidence that we shall later discuss. Here, indeed,
is the central problem of alchemy.

With the fraudulent alchemist we are on much surer
ground. The swindles which could be worked were so sim-

ple and attractive that it is not surprising that papal bulls
and civil laws had to be enacted against the so-called Multipliers of Metals. The numerous swindlers who practiced this
livelihood did not find any need for lengthy processes. They
professed to possess the stone or to have a simple and quick
recipe for making it. It is notable that even today gold, in
the form of the brick and the gold mine, is the subject of
the standard swindles—pretended transmutation of base
metals to gold was their medieval representative.

The usual procedure was to interest a wealthy man,
usually a priest (the clergy are still favorite subjects of the
swindler's art), and to use the timeless technique of the
charlatan to lead him to seek a demonstration. The trickster
provided himself beforehand with some gold and silver. A
furnace was set up, a crucible and mercury were purchased,
the crucible was filled with mercury, and the precious powder—probably a little chalk or red lead—was poured into it.
Meanwhile some genuine gold or silver had been conveyed
into the interior of a piece of charcoal or a hollow in the
end of a stirring-rod, and kept in place with blackened wax.
The furnace was heated; the prepared charcoal was conveyed to its place above the crucible, or else the rod was
brought into use. The wax melted and the precious metal
fell into the mercury; as the heat grew greater the mercury
volatilized and left the melted gold or silver in the crucible.
What more could be needed as proof? The dupe easily
parted with large sums for the purchase of quantities of
laboratory materials and mercury, or paid a large sum for
the recipe for making the stone, whereupon the fraudulent
alchemist was seen no more.

The signs of a fraudulent alchemist were his itinerant life
and his readiness to talk. The true alchemists seem to have

spent long years in their laboratories, and, if we believe those who claim to have had experience, they took great care to dispose of their gold in secrecy, desiring nothing less than to be known as alchemists.

An amusing enough chapter of such swindles could be compiled, but it is alchemy that we wish to understand and not the vagaries of the medieval trickster. The reader who would pursue these further may read Chaucer's *Canon's Yeoman's Tale* or that delightful play, Ben Jonson's *Alchemist*. Both authors were very well acquainted with alchemy, and doubtless had firsthand knowledge of many alchemists or pretenders to alchemy.

IX

Alchemy in the Fourteenth Century

THE books which were available in the thirteenth
century were translations or paraphrases from the
Arabic and for the most part were both obscure
and mixed with a good deal of unnecessary jargon. The
best works available were those of Geber (pp. 91-94) whose
Summa perfectionis and other works were Latin versions
of the best practical ideas of Arab alchemy and contained
the standard chemical preparations and methods used by
later authors. But these were works far superior to most
of the Arab translations, which are commonly tedious and
confused.

The mind of the late thirteenth and the fourteenth centuries was exceedingly clear and rational and had the scholastic habit of liking a subject to be presented in a systematic
and logical way. So, whether for this reason or others, we
find in the fourteenth century a series of alchemical books
written by European authors and of quite a different tone
from the Arabic translations.

The usual difficulty is found in discovering whether these
books had any connection with the authors whose names
are attached to them. Thus Arnald de Villanova and Ramón
Lull are the names affixed to the most influential alchemical

writings of the early fourteenth century; but it is pretty certain that only a few of the texts attributed to Arnald de Villanova were written by him, and it is probable that none were written by Ramón Lull. None the less, these writings are of the early fourteenth century and contemporaries or their immediate posterity believed them to be genuine. We shall speak, therefore, of Arnald or Lull and intend by the names the authors of the treatises later accepted as theirs.

The most interesting group of writings of this period are those to which is attached the name of Ramón Lull. Many of them are dated in the texts as c. 1330. As Lull died in 1315 they cannot be directly from his pen, but probably emanate from his followers. Ramón Lull was a remarkable man. He was born in Majorca, between 1232 and 1236, and led a dissipated court life till his conversion in 1266. He determined to devote himself to the conversion of the Moslems and spent nine years in learning Arabic. He believed that the conversion of Islam could be achieved by the refuting of its philosophers, especially Averroes. Lull's views were a rationalistic mysticism. He believed that theology and philosophy were one, and that the profoundest truths of revelation, such as the existence of three Persons in one God, could be proved by reason. He invented an extraordinary scheme for a mechanical system of logic whereby the premises of theological propositions could be mechanically arranged in various orders by a machine or by complicated tables and so prove themselves true. He was martyred by the Saracens in 1315, but he has never been canonized as a saint, presumably because his views about faith and reason are seriously in error.

His followers formed an influential school of philosophy, and there is no reason to doubt that alchemical works were

110

written by some of them and attributed to the master. Perhaps the most important alchemical work of the middle ages is the *Testament of Lullius*, divided into three parts, the *Theoretical*, the *Practical*, and the *Codicil*. These books are a systematic account of a theory and practice of alchemy which appears to have been new, though it may well be found to have Arabic antecedents.

The Lullian works are distinguished by numbers of tables (Plate I) in which the principles, materials, and operations of alchemy are symbolized by letters of the alphabet and various processes indicated by further combinations of these letters—a method which does not make for easy reading. But the most interesting thing about them is that, unlike most alchemical works, they are not a mosaic of quotations from earlier authors but a consistent logical work, and that they have very little in them which is allegorical or deliberately obscure. The picturesque array of green lions and tail-eating dragons, the red man and his white wife, the king and queen, the gold and silver trees play in them but the smallest part.

Lull's doctrine stipulates that the thing which God created was what he calls "argent vive" (*argentum vivum*, quicksilver, mercury), and that this original matter gave rise to all other things. The finest part formed the bodies of the angels, a less fine part the heavenly spheres, stars and planets, and the coarsest formed the terrestrial bodies.[1] But in the terrestrial bodies part of this "argent vive" became the four elements earth, water, air, and fire, but a part remained as a fifth element, the *quintessence*. Thus, in every body there was some stuff akin to the heavenly bodies,

[1] Cf. quotation from Synesius, p. 15.

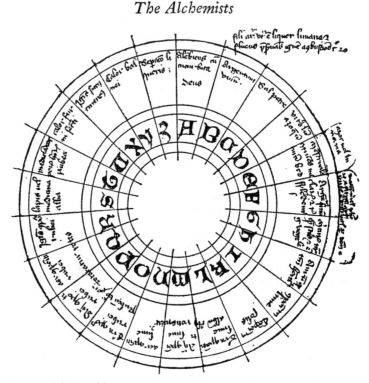

Fig. 13.—Table of letters and materials of the Art, from a fifteenth-century manuscript of the Lullian treatises. (Courtesy of Mr. D. I. Duveen.)

and it was through this material that the heavenly bodies could bring about the changes of generation and corruption. The activity of the body abode in the quintessence, and alchemy was a process dealing with this fifth element and multiplying the activity in it. This theory, developed in a hundred or so pages of text, differs from most alchemical works in not being deliberately mysterious. The *Practical* part is likewise fairly clearly described, but not so that one can pick out any short, easily intelligible portion for discussion.

There is, however, a very clearly phrased book, attributed to Lull and dated 1330, which gives numerous intelligible accounts of chemical or alchemical processes. This is *The Experiments of Raymond Lully of Majorca the most learned philosopher, wherein the operations of the true Chymicall Philosophy are plainly delivered.* The Latin text was edited in 1572 by one Michael Toxites, together with some of Lull's minor works. An English translation was made, but not published, by one William Atherton in 1558, and from this [2] I will quote one of several recipes for making the philosopher's stone. This recipe, though unusually clear, is in many ways typical. It starts with gold, silver, and a "philosopher's mercury," a volatile distilled liquid which here seems to be nitric acid; and it follows approximately the more elaborately and less clearly set out processes of pages 142-143. The reader may be reminded that the sign ☉ (Sol) represents gold, ☽ (Luna) silver, and ☿ mercury:

THE 33RD EXPERIMENT OF SOL

Take aqua fortis with his form, as I have taught you before, and in it dissolve 3 ounces of Lune; then purify it 20 days, then take 3 oz. of ☉ and dissolve it in 18 oz. of the same aqua fortis with his form in which yet 4 oz. of the fixed salt of Urine ought to be first dissolved, as you have it in his Experiment. Then putrefy these 2 bodies by themselves severally for 20 natural days. Then exanimate [i.e., remove the spirit of] them both severally by themselves, as well Lune as ☉, even to the rule delivered you before. Now when every one shall be exanimate by themselves and their quickened waters shall be severally kept by themselves and also the earth shall yield no more smoke, then

[2] MS Ashmole 1508, contained in the Bodleian Library.

shall there be a sign that the ☉ and ☽ do suffer eclipse. Beate then the earth of either of them, and likewise mingle them then in a little glass ball strongly luted. Put him to a fire of reverberation 24 hours.

Then take it out and give it first the water of ☽ quickened and rectified first 7 times by ashes. And when it shall have drunke up all his Water by little and little, in the same order as you had it before in the other experiments, then give it the Water of ☉, without any rectification, by little and little after the order which you kept in imbibing that earth with water of ☽. Then you shall ferment it in this manner. Take one part of ☉ and 3 parts of ☿ and one part of the medicine, that is to say as much as there was gold. Put it all together in a glass vessel upon warm ashes, and in a short time it shall be turned into powder. Then shall you incere it with the 3rd oil of ☉. Now when it is all very well cerated and brought into the forme of oyle, project one part thereof upon 100 of mercury: and it shall be all turned into medicine. Of which take againe one part and project upon 500 parts. It and turn the mercury itself into ☉ better and purer than natural gold.

As the language of this is somewhat different from that of modern chemistry I will explain it in the language of the latter:

Three ounces of silver is dissolved in nitric acid. Three ounces of gold is dissolved in 18 ounces of nitric acid in which was dissolved four ounces of a product chiefly consisting of common salt. Solutions (a) of silver nitrate and (b) of gold chloride are obtained. Both solutions are left for 20 days, then distilled to dryness. The distillates consisting (a) of fairly strong nitric acid and (b) of nitric acid containing some chlorine are preserved. The dry salts (earth) are heated till they give no more fumes. They are then ground and mixed and heated in a glass vessel. The result would be a mixture of finely divided metallic gold and silver chloride, together with some salt derived from the "fixed salt of urine."

The nitric acid (a) is distilled seven times on the gentle heat of ashes. This is then added to the solid mixture (which is probably heated so as to evaporate some liquid), then the distillate (b) is added, the result probably being a mixture of silver chloride, gold chloride and the free metals (but the directions are not sufficient to determine this). Then one part of this mixture is added to a mixture of 3 parts of mercury and one of gold. By warming it a powder is obtained—gold amalgam with admixture of gold and silver salts. It is now made into a paste with "oil of gold," which usually means the solution of gold chloride in hydrochloric and nitric acids. Much of the metal will dissolve and form a thick oily solution or paste of mercury, gold and silver salts. By throwing one part of this on 100 parts of mercury a very weak gold amalgam would result, which when added to 500 parts of mercury would have no effect whatever.

To such an anticlimax come all attempts to give a chemical interpretation to recipes for transmutation. The above is a recipe much more intelligible than most, describing chemical processes that can be identified, but ending up with the assertion of a transmutation that could not possibly result from the use of this material.

What is the explanation of this general phenomenon— that alchemists describe what, according to modern science, could not have occurred? The alchemist, if we could tax him with this, might, of course, simply ask us whether *we* had tried the recipe—and, of course, no modern scientist has done so. Before anybody will contemplate spending weeks or months on a piece of work they must have some hope that something will come of it. Thus we often hear stories of people who have caught sight of Noah's Ark on top of Mount Ararat; but no expedition is formed to study this remarkable relic. The occurrence of a flood, seventeen thousand feet in depth, which would have been needed to float it there, does not seem sufficiently probable to induce any one to spend time and money on investigating the question. So with alchemy. These recipes could be tested, but nobody believes in them enough to think it worth doing.

If we assume that these recipes will not avail to make gold, what are we to think? Did the alchemists invent recipes which they thought would work, but had not fully tested? This is not at all unlikely. Much later, in the sixteenth century, we find books of illustrations of machines that could not have worked and were evidently never made, and accounts of physical experiments which could not have succeeded. It is, therefore, not in the least unlikely that the alchemists, on the basis of some signs of success (e.g., the whitening or yellowing of some metal), designed methods which they genuinely supposed would be successful, but which the frequent accidents of the laboratory prevented them from bringing to an end.

Another explanation of these recipes is given by those who hold that they were not chemical at all, but represented in a symbolic fashion operations purely mental—that the true alchemy was not concerned with chemicals and apparatus but was a sort of mystical operation. Of this view I shall have more to say in Chapter XIV.

Perhaps the most important part of the Lullian treatises is not the theory and practice of gold-making, but their description of the preparation of "quintessences." We have seen already how in the belief of many alchemical authors there was a subtle spirit diffused throughout the world. That spirit is described as a *quinta essentia*, a fifth being, over and above the four elements, and, as we have seen, it was believed to exist in all earthly bodies and to be their active principle. It was clear to the Lullian alchemists that, if this active principle could be extracted from a body, it should be a far more active reagent than the body itself. It is true that alchemists had for a thousand years been extracting volatile "spirits" from bodies by distillation and had not accomplished very much. Some of these "spirits" proved uninteresting, being mainly water, some were powerful reagents, such as the mineral acids and ammonia; but an entirely new interest appeared with the discovery of alcohol.

Wine had been distilled and an *aqua ardens*, a "water which would burn," had been extracted from it at a date which is disputed, but which is at least a century and probably several centuries before the Lullian treatises were written. Yet this distillate does not seem to have attracted much attention and was, so to speak, a chemical curiosity. In the thirteenth century it began to be used as a medicine, and by 1288 it was evidently in general use. Thus the Dominican provincial chapter at Rimini in that year forbade the brethren to possess the "instruments by which they make the water called *aqua vitae*." Arnald de Villanova describes it in 1309-12 and extols its curative virtues.

But the works attributed to Ramón Lull seem to be the first to treat this *aqua vitae* as an impure form of the quintessence, the active spirit. He gives many recipes for making it. Good wine is first distilled, through "copper side tubes," and four times redistilled and kept carefully stoppered. Its strength was tested by a quite effective method. Some of the spirit was poured on a pile of sugar or a piece of cloth and set alight. If the spirit contained much water, the cloth or sugar was left so damp by the time the spirit had burned off that it would not catch fire. But if the spirit, when dropped on cloth or sugar and set alight, would ignite them, it was known to be strong enough. The spirit could be further strengthened by distilling it with calcined tartar (mainly anhydrous potassium carbonate) which absorbed some of the water. The result was probably alcohol of 90-95% strength. But, characteristically, the end product of this clearly described chemical process was regarded only as a "burning water," *aqua ardens*, and it was not considered to become the real quintessence until it had been "circulated."

To this end a large quantity of fermenting matter, such as dung, was heaped up to form a hotbed, the "aqua ardens" was placed in a "pelican" or similar vessel (Plate II) and sealed up. The vessel was then half embedded in the dung-bed. The warmth volatilized the alcohol, which condensed on the projecting part of the pelican and ran back again, so setting up a circulation (as in the modern reflux condenser).

All this is clear; but the result was something that science knows nothing about. The *aqua ardens* was said to separate into two layers, a lower turbid layer, which was to be rejected, and an upper layer, clear and sky-blue, which was the quintessence. When the vessel was opened, a wonderful fragrance exhaled so that birds were attracted to it and people would come crowding into the house. We cannot think, of course, of any possible change that "circulation" could bring about in alcohol. We must class this part of the story with the rest of the inexplicable marvels of alchemy, yet noting the analogy with the great world in which the original "argent vive" separated into the blue of heaven and the dull, turbid mass of earth.

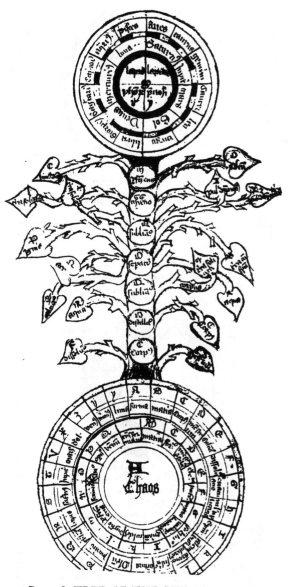

PLATE I. TREE OF THE PHILOSOPHERS
(From a fifteenth-century manuscript of the
Lullian treatises. Courtesy of Mr. D. I. Duveen)

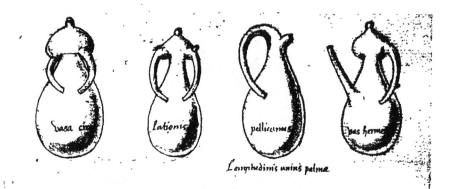

PLATE II. HERMETIC VESSELS
(From a sixteenth-century manuscript)

PLATE III. THE ALCHEMICAL DRAGON

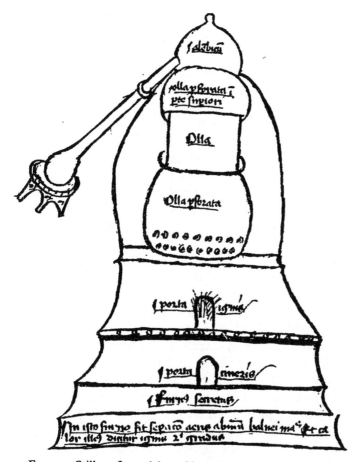

FIG. 14.—Still as figured in a fifteenth-century manuscript. The alembic is heated in an air bath, over a charcoal furnace. (Courtesy of Mr. D. I. Duveen.)

Nevertheless, the result of this circulation, even if it did not come up to the standard of the recipe and was only a specimen of nearly pure alcohol, was really a marvel. Its effect on the human organism was very obvious. The reputation of brandy as a restorative remains today. The effect of it on the chilly limbs and failing powers of the aged was most impressive to the men of the time. So we find it in succeeding centuries used as a medicine against old age. Its power of preserving organic matter from putrefaction probably also helped to support the idea that it would preserve the human body "till the term fixed by God." Moreover, the notion that this medicine was the very quintessence, the active spirit of the world, gave reason for the strongest presumption that it would prove to be the most perfect of medicines. As quintessence it was the link between our earthly bodies and the heavenly bodies, and could transmit to us their beneficent influence.

But this alcohol also had a chemical property new to the world, for it was the first liquid solvent for many organic compounds insoluble in water—such as fats, resins, and essential oils. It was, therefore, the first known liquid that could extract the volatile aromatic substances from plants. These volatile essential oils with their aromatic odors and fiery tastes seemed obviously to be the "quintessence" of the plant. Every plant had its stars—"There is no herbe but hath his starre and it shoots him with his beame and saith to him 'Grow!'" The star's celestial influence entered into the celestial part of the plant, its quintessence, and *that quintessence could be extracted by means of the quintessence of wine.* So the Lullian treatises and other fourteenth century works on the same subject, deal, amongst other matters, with the extraction of the quintessences of plants by solution or distillation with alcohol. The result, of course, only differed from a modern liqueur in that it was not sweet, and honey or sugar may well have been added to make the medicine more palatable.

Now the oldest liqueur that we know of is Benedictine, invented in 1510 by Dom Bernardo Vincelli, who is known to have been a chemical worker. This admirable liqueur is made by the extraction and distillation of a great number of herbs with alcohol, and it is reasonable to suppose that it was designed as a quintessence extracted from a great variety of plants, containing in it all the heavenly influences that had made them grow. In it and its kind we may find not the least of the alchemists' achievements.

From the time of the Lullian treatises this "quintessence of wine" or "circulatum" was an important substance both in medicine and in alchemy. Needless to say, the world also soon found out that it was not only a good medicine but a good drink. Michael Savonarola, grandfather of Jerome Savonarola, would-be reformer, wrote a book (c. 1430) on the subject, wherein he tells us of a friend who drank his eight ounces a day and swore it was the only reason he had lived to be eighty. But Savonarola on the whole disapproves of treating this wonderful medicine so lightly. We should take it, he says, like another sacrament and not merely as a remedy for a hang-over. Half a century later, when printing had come into use, there was a flood of "distillation-books" telling how to make all manner of distilled liquors, but still treating them as medicinal; the general use of distilled liquors for convivial purposes seems to date only from the last quarter of the sixteenth century.

The Lullian treatises are by no means the only important

early medieval alchemical texts. Thus the *Mirror of Alchemy*, attributed to Roger Bacon, and the *Straight Path*, ascribed to Albertus Magnus, pursue a somewhat different line of thought from the Lullian writings. But space forbids the discussion of many of these texts. In my own judgment, the Lullian view of alchemy was historically the most significant and fruitful, both in the developing of the alchemical natural philosophy and in leading, through the work of Paracelsus and his followers, to chemistry.

X

The English Alchemists

IT is not possible in a book of this kind to trace the story of medieval alchemy in the many countries in which it flourished. The history of English alchemy is however fairly well known and though it is not so long as that of Italian or French alchemy, it serves well to indicate some of the general features. We have two main sources of knowledge, the public records and the writings of the alchemists. The former are scanty but accurate, the latter voluminous but of more doubtful veracity. Both sources indicate that English alchemy began in the first quarter of the fourteenth century.

The public records are concerned chiefly with the practical man, honest or fraudulent, who was trying to make silver or gold in quantity, whereas the writings of the alchemists give us the reflections and theorizing of what we may call the alchemical philosophers, the men who were known to later generations as the great masters of the Art.

The first record of the former type is in the *Patent Rolls* [1] for 1329. Johannes de Rous and Willielmus de Dalby, who were believed "to know how to make silver metal and previously to have made metal of this kind" were summoned

[1] Thomas Rymer, *Foedera*. London, 1727, Vol. IV, p. 384.

to come before King Edward III with their instruments and other things necessary to demonstrate an art which clearly promised so much good to the country. No more is heard of the matter and, indeed, alchemy vanishes from the English scene until the end of the century. By that time it was clearly a scandal.

Geoffrey Chaucer in the thirteen-nineties fiercely satirizes the alchemical swindlers in his *Canon's Yeoman's Tale*, and in such terms as make us feel sure that he was very well acquainted with the subject, and probably to his cost.

The fact that alchemy was then a serious problem is further witnessed by the promulgation of a *Statute* in 1403-4 against the multiplying of metals, in the following terms:

It is ordained and stablished, That none from henceforth shall use to multiply Gold or Silver nor use the Craft of Multiplication: And if any the same do, and be thereof attaint, that he incur the pain of Felony in this case.[2]

The penalty for felony was death and forfeiture of goods. The statute certainly did not put an end to alchemy, for more than twenty persons during the fifteenth century obtained licenses from the King to practice it. Here is an example from the *Patent Roll* of 24 Henry VIth (1445):

Know that since William Hurteles, Alexander Worsley, Thomas Bolton and George Hornby have signified to us that inasmuch as they wish to work by the art of philosophy on certain materials, namely to transform the imperfect metals from their own nature, and then by the said art to transubstantiate them into perfect gold and silver according to every kind of test and examination, just as any gold and silver in its ore is to

[2] *The Statutes of the Realm*, ed. A. Luders *et al.* London, 1816, Vol. II, p. 144.

be waited for as it grows and is to be hardened, they say nevertheless that certain malevolent and malignant persons suppose
them to be operating by an illicit art, so that they are able to
hinder and disturb them by their disapproval of the said art,
We considering the aforesaid matters and wishing to know the
conclusion of the said work, of our special grace have granted
and given licence to the same William, Alexander, Thomas and
George that they may work on and test the aforesaid art
without hindrance from us or any of our officers whomsoever.
Provided always that to do so does not offend against our law.

Notice that the petitioners allege that they are operating
in a natural and lawful manner. The *Statute* was aimed at the
deliberate frauds and the genuine experimenter had no difficulty in obtaining a license, especially as the Kings of England in the time of the Wars of the Roses were always
extremely short of money and accordingly ready to en-
courage the makers of gold. Moreover, as becomes apparent in the quotation that follows, the alchemists were
ready to pay for their license.

There are a few references to these licenses in the writings of the alchemists. One of these is in the *Liber Patris
Sapientiae*, which appears to be of the latter part of the fifteenth century.

Therefore keep close of thy tongue and of thy hand,
From the officers and governors of the land;
And from other men that they of thy craft nothing know
For in witness thereof they will thee hang and draw.

And thereof the people will thee at Sessions indict,
And great treason against thee they will write
Without that the King's grace be to thee more,
Thou shalt for ever in this world be forlore.

125

Also without thou be sure of another thing,
To purchase the license of thy King:
For all manner of doubts thee shall betide,
The better thou mayst work, and both go and ride.[3]

Despite the possibility of obtaining such licenses, many practiced multiplication of metals contrary to the law. Thus in 1452 the King empowered three commissioners to arrest multipliers of metals. However, the most interesting record is that of the trial of William Morton in the year 1419. He was indicted for having said that he could make a red powder, called elixir, which, projected on any "red metal" such as copper, bronze, or brass, would turn it into gold, and, if projected on a white metal, would turn it into silver. He joined forces with a monk of the priory of Hatfield Peverel and worked with mercurial powder, charcoal, vermilion, verdigris, nitre, alkali, glass gall,[4] vitriol, arsenic, and the like. With them he made a black mass, congealed in a round vessel, and asserted to Joan, Countess of Hereford, and two justices of the peace that he could in ten weeks make this into the elixir, by means of which he could turn red and white metals into true gold and silver from which the King's money could be coined[5] and made. He was found guilty, and after a short spell in Colchester gaol was pardoned.

It seems, then, that the Statute against the multiplication of metals, though it was not a dead letter in the fifteenth century, was very leniently administered. Licenses con-

[3] Elias Ashmole, *Theatrum chemicum Britannicum.* London, 1652, p. 196. (Spelling is to some extent modernized.)
[4] A salty scum thrown off by glass during fusion of its ingredients.
[5] It was the custom to bring gold to the mints for coining.

tinued to be issued at lengthening intervals up to the early years of the sixteenth century.

Such then is the view of alchemy that we obtain through the State papers. A very different picture, though not an inconsistent one appears in the works of the alchemical writers.

The earliest English alchemical writings, like the earliest public records, date from the early fourteenth century. The works of John Dastin and the *Ycocedron* of Walter Odington belong to this period. Moreover, some of the treatises attributed to Lull assert their origin to be England and their date about 1330. We cannot be sure, however, that they were written so early, though only negative evidence can be brought against this possibility. These Lullian treatises are of particular interest to English-speaking readers, because they are the chief source of the school of English alchemists, about whom we have a fair knowledge, almost all, however, derived from their own works. Most of these are printed in the collection of English alchemical poetry made by Elias Ashmole in the seventeenth century, the *Theatrum chemicum Britannicum*. This rare book should be reprinted as it is a prime source of alchemical information. The story of English alchemy that we gather from these writers suffers, of course, from all the errors to which tradition was subject in an age which had little idea of historical accuracy. Nevertheless, it certainly tells us of many real people and gives an interesting picture of the late medieval world of alchemy.

The first of the English alchemists was supposed to be Abbot Cremer of Westminster. His *Testament*, which contains his story, first appeared in print in the *Tripus Aureus* of Michael Maier (1618), but, as we shall see, there is very little evidence of his real existence. Here is his story, as told in an English manuscript (Ashmole, 1415):

And I, being a very earnest follower of this art and faculty, was most wonderfully holden back by means of matter to me shewed and declared very obscurely in diverse and sundry books, which I read and exercised according to the instructions thereof by the space of 30 years to my great cost. and loss of labour. And the more I read the more I erred, until at the last by the Divine providence I came into Italy where it did please the most high and mighty God to bring me in company with a man no less endowed with dignity than all kind of learning, whose name was Raymond, in whose company and fellowship I stayed long to the end that he should open some part of this great mystery unto me; and furthermore I handled him with many entreaties that he came with me into this Island and stayed with me two yeares, in which space of time I got and obtained all the work.

And furthermore I brought this excellent man unto the sight of the most famous K. Edward, of whom he was most worthily received and kindly entertained; and being there with many promises, covenants and agreements moved and persuaded by the king, he was contented by the sufferance of God with his art to enrich the king, upon this only condition that the King in his own person should fight against the Turk, the enemies of God, and that he should bestow somewhat of the house of our Lord, and nothing at all in pride or warring against Christians, but (oh for sorrow) his promise was broken and violated by the King. Then that holy man sore afflicted in spirit, and secrets of his heart fled hence beyond the sea in most lamentable and miserable manner which burneth mine heart not a little. I do most heartily desire to be with him daily in my body, for the beholding of his daily life and integrity of his manners would soon bring most obstinate sinners to repentance. Oh happy and blessed Raymond, I will pour out for thee prayers unto the most high and mighty God, and likewise my breathren will do the same.[6]

The King in question is evidently Edward III, and the name Raymond refers to Ramón Lull. The breaking of the King's promise was presumably the making of war with France in 1337. There are other sources of this story, for Ashmole, in the preface to his *Theatrum chemicum Britannicum*, gives some details that do not appear in the *Testament*. It is true (p. 123) that Edward III sought to obtain silver by alchemical methods at this period, but there are obvious reasons to disbelieve the history of Lull and Cremer.

First of all, Lull died in 1315, some twenty years before these events; also, there is no record of a Cremer among the Abbots of Westminster, and, most significantly, he is not mentioned by Ripley or Norton. Yet the story must be fairly old, for the copyist of MS Ashmole 1415 tells us that it was taken from a "very ancient" parchment copy, which must therefore have been not much later than the year 1400. Ashmole notes on this manuscript that "Dr. Dee[7] concevyth this was written by the Prior of Ramsey," and the name of Cremer appears nowhere, except in a note added by Ashmole. Some of the Lullian treatises are dated in the text as having been written in England about 1330. This is quite possibly true, but their real author was not Lull. It is quite likely, however, that the Cremer legend was made up to account for the apparent presence of Lull in England.

[6] This is the version of the *Testament* contained in MS Ashmole 1415, but it differs from that printed by Maier.

[7] John Dee, astrologer, sorcerer, alchemist, mathematician (1527-1608).

That alchemy was flourishing in fourteenth-century England is obvious from Chaucer's *Canon's Yeoman's Tale*, written about 1390, but in the fifteenth century, we find extensive records of the English alchemists and numerous works in verse and prose, both Latin and vernacular. We could compile a very long list of names of students of alchemy belonging to the fifteenth and sixteenth centuries in England, but there is a clear distinction to be drawn between those reputed to have been masters, i.e., to have made the stone, and those who merely wrote about it or practiced unsuccessfully.

Among these masters the first place is taken by Sir George Ripley, Canon regular of Bridlington in Yorkshire, who learned the Art in Italy, as may be seen from his works, and seems to have been working between 1450 and 1490. At about the same period lived Thomas Norton of Bristol, whose *Ordinall* was begun in 1477. He does not seem to have learned the Art from Ripley, as appears on internal evidence and from the testimony of his great-grandson, Samuel Norton, also an alchemist. Ripley apparently handed on the Art to a "Canon of Lichfield," who gave it to Thomas Daulton, who died before 1471. Ripley seems to have handed on the Art too, through an unknown intermediary, to William Holleweye, alias Gibbs, elected Prior of Bath Abbey in 1525, who transmitted the secret to Thomas Charnock. The latter had another master, whom he refers to as I. S. or as Sir James, and who was a "Priest of the close of Salisbury." This I. S. had no master, "God having put the secret into his head" as he was lying in bed. Thomas Norton also mentions three masters of the Art, one of whom Ashmole records to be of the name of Crosby, and of whom no more is known.

The English Alchemists

There were, of course, very numerous other alchemists at this period, honest and otherwise, but those we have mentioned were reputed as honest masters of the Art, and they form a compact group, employing, as far as we can see, similar methods.

The date of Ripley's birth is unknown, but he was certainly writing on alchemy between 1450 and 1476, and he died in 1490. He was apparently of a good Northern family. He was a canon regular of the famous Augustinian Priory of Bridlington in Yorkshire, a great home of learning where William de Newbridge and other historians had worked and where the last English saint to be canonized before the Reformation, John de Tweng, had been prior until his death in 1379. The last prior, W. Wold, gave up his life rather than betray his faith. Ripley was, however, exempted from the observances of his order and may be thought to have been more a scholar than a monk.

The preliminaries to his *Compound of Alchemy* show that he learned the Art in Italy, at that time a great center of the less orthodox sciences and philosophies, and also that he had stayed at Louvain and elsewhere in "farre countries." Elias Ashmole (1652) tells us in a manuscript note that he visited the Isle of Rhodes and resided there for some time with the Knights of the Order of St. John of Jerusalem. "An acquaintance of mine hath in his custody certain private observations of an English gentleman of good quality and credit who in his travells abroade, observes (amongst other things) that in the Isle of Malta he saw a Record which declared that this Sir George Ripley gave yearely to those Knights of Rhodes £100,000 towards maintaining the war (then on foot) against the Turks." We need not give too much attention, however, to such secondhand evidence. At the end of his life he returned to England and became a Carmelite and for the two years before his death in 1490, lived as a hermit at St. Botolph's near Boston in Lincolnshire.

Thomas Norton came of a Bristol family of some importance. He was a privy councillor and we may conclude him to have been much more wealthy than most of the alchemists. From his *Ordinall* we can discover that he began the study of alchemy at an early age, that he rode more than a hundred miles to meet his master, and in forty days spent with him was taught all the secrets of alchemy, attaining the preparation of the elixir of gold at the age of 28. The elixir was stolen from him, as also was the elixir of life, which he subsequently prepared. His *Ordinall* is a long rambling poem in which he tells us a great deal about the ways of alchemists, though not very much that is definite concerning the work.

One of the alchemists he mentions, Thomas Daulton, is an interesting figure. He does not appear to have left any writings, for although I have met *Daulton's Degrees of Fire* as a title, the text that followed it was the seventh chapter of Norton's *Ordinall*. But the details that Norton gives concerning him are interesting.

Thomas Daulton, he tells us, was a good man, who had a great store of the red medicine. One of Edward IV's squires, Thomas Herbert, took him from an abbey in Gloucestershire and brought him to the King. Daulton had been "clerke" to Sir John Delves, who was also "squire in confidence" to the King. Delves, despite the fact that he had been sworn to secrecy, told the King that Daulton had made him a thousand pounds of good gold in less than half

a day. Daulton then told the King he had thrown the medicine into a lake to avoid a recurrence of the troubles it ha
already brought him. The King released him, but Thor
Herbert lay in wait for him and carried him off to Glouce
ter Castle and then to his own seat at Troy in Monmou
shire, where he kept him four years. Daulton refused
reveal the secret even under threat of immediate executio
and died soon after his release. Norton tells us truly that Herbert died soon after and Delves lost his head at the battle of
Tewkesbury.

It is quite clear from the usual historical sources that Sir
John Delves and Lord Herbert of Troy were real persons,
and since the former was killed at the battle of Tewkesbury
in 1471, Daulton must have died about that time. Norton is
here writing about events not ten years old, and we must
presume that they are at least based upon facts.

About Thomas Charnock we know a good deal more,[8]
because his own work was partly autobiographical and because Elias Ashmole collected a good many interesting an-
notations from books and MSS he had owned, some of
which were then extant. From these sources we can collect
a story which insofar as it concerns Charnock himself may
be presumed to be true, though less credence should be
given to what Charnock tells us about others, e.g., concerning the Prior of Bath.

Thomas Charnock was born in 1524 or 1526. He was a
man of small education, able to read and write and with a
turn for direct and vigorous English verse, but with a rather
scanty knowledge of Latin. When he was about twenty-one
or twenty-three he began to travel all over England trying

[8] F. Sherwood Taylor, "Thomas Charnock," *Ambix*, Vol. II, pp. 148ff.

to find someone to teach him the secret. When he was
twenty-eight he was given the secret by "I. S.," the priest
of Salisbury (p. 130), who, believing himself to be near
death, also gave Charnock the "Work" he had begun. But
Charnock does not seem to have learned the secret fully,
for he was instructed again by "a monke of Bath which of
that house was Pryor."

If we credit Charnock, this must be the last Prior of Bath
Abbey, William Holleweye, sometimes known as Gibbs,
who, on the occasion of the dissolution of the monasteries,
surrendered the abbey to the Crown in 1525 and received
a pension of eighty pounds a year. Charnock knew him
nearly thirty years later when he must have been an old
man. The Prior told him, incidentally, that he used the water
of the famous hot spring at Bath to give the gentle heat required for the work. The Prior had owned the red stone
and, when the Abbey was suppressed, he hid it in a wall;
but when he returned a few days later and sought for it,
it was gone. Thereupon he lost his reason and wandered
about the country. When Charnock knew him he had become blind and had to be led about by a boy. Ashmole
corroborates this story by a manuscript note in his an-
notated copy of his *Theatrum chemicum Britannicum*
(1652):

Shortly after the dissolucon of Bath Abbey, upon the pulling
downe some of the Walls, there was a Glasse found in a Wall
full of Red Tincture, which being flung away to a dunghill,
forthwith it coloured it, exceeding red. This dunghill (or
Rubish) was after fetched away by Boate by Bathwicke men
and layd in Bathwicke field, and in the places where it was
spread, for a long tyme after, the Corne grew wonderfully
ranke, thick, and high: insomuch as it was there look'd upon

134

as a wonder.[9] This Belcher and Foster (2 Shoomakers of Bath, who dyed about 20 yeares since) can very well remember; as also one called Old Anthony, a Butcher who dyed about 12 yeares since.

This Relacon I recd: from Mr. Rich: Wakeman Towne Clearke of Bath; (who hath often heerd the said Old Anthony tell this story) in Michaelmas Tearme 1651.

The Prior told Charnock the whole secret after binding him with an oath of great solemnity, and it is notable that he was able to tell it in "three or four words." Presumably this remarkable brevity was only possible because Charnock had already much knowledge of the Art.

Charnock busied himself again upon the "work" that his first master had left him, but his apparatus caught on fire and he lost it. He made two more attempts, but just as the work began to promise success, a gentleman who "owed him much malice" caused him to be pressed (conscripted) for the army that was sent over to defend Calais (1557-58). He must have been a man of little means, otherwise he could have bought himself off; but as he complains much about the expense of the work in glasses, firing, and the like, it may be that his credit was exhausted. Be this as it may, he fell into a fury and smashed all his apparatus with a great hatchet. The siege of Calais seems to have afforded him some leisure, in which he wrote a most amusing poem. Only after this, on his return to England, does he seem to have settled down to the long work of making the stone, which occupied most of his life. Whether he had any success is not clear: in 1574 he made a note which indicates that he believed himself to have made the white stone; but in 1576

[9] The philosopher's stone was believed to perfect every being in its own kind; hence its supposed effect upon the corn.

he was still apparently not at an end, and in 1581 he died.

The interesting things about Charnock are, first, the fact that we have some knowledge of him apart from his own work, and secondly, the humor that shines through his crude verse.

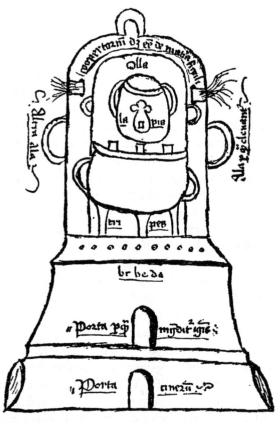

Fig. 15.—The apparatus for making the stone, from a fifteenth-century manuscript of the Lullian treatises. (Courtesy of Mr. D. I. Duveen.)

136

He tells us of the difficulty of keeping the work secret, so that one should not be known as an alchemist. The craftsmen who made the instruments were curious about them. The potter, for example, had to be satisfied; it was difficult to get him to make what he had never seen, so the alchemist was to stand over him to show him what to do. But then he was sure to be curious about the vessels, and the alchemist should say that his father was somewhat blind and he was going to distil a water to cure his eyes. Then the joiner was required to make an outer wooden cover for the apparatus which could be locked so that it could not be seen or disturbed. So the joiner would also wonder what he was making, and the alchemist should "laugh and say is it a burrow for a fox, although it be made sure with keys and lock."

The glassmaker presented a greater problem, for there were few in England. Charnock tells us of one at Chiddingfold in Sussex who had to be approached "in humble wise" and requested to blow glasses of the form desired. A journey from Somerset, where Charnock lived, to Chiddingfold was no light enterprise in the 1550's.

His neighbors were evidently put off in the same humorous fashion. A hundred years after Charnock's death a clergyman, named Pascal, heard that a roll of manuscript had been found in a wall of Charnock's house at Combwich. He went there to make inquiries and thus reported:

I was also since my last at Mr. Charnocks house in Comage, where the Roll was found, and saw the place where it was hid. I saw a little roome, and contrivance he had for keeping his worke, and found it ingeniously ordered, so as to prevent a like accident to that which befell him New Years day 1555, and this pretty place joyning as a closet to his Chamber was to make a Servant needless, and the work of giving attendance

137

FIG. 16.—The painting on the door of Charnock's laboratory.

more easy to himself. I have also a litle iron instrument found there which he made use of about his Fire. I saw on the dore of his little Athanor-room (if I may so call it) drawne by his owne hand, with coarse colours and worke but ingeniously, an Embleme of the Worke, at which I gave some guesses, and soe about the walls in his Chamber, I think there was in all 5 panes of his works, all somewhat differing from each other, some very obscure and almost worne out.

They told me that people had been unwilling to dwell in that house, because reputed troublesome, I presume from some traditionall stories, of this person, who was looked on by his Neighbours as no better than a Conjuror. As I was taking Horse to come home from this pleasant entertainment, I see a pretty antient man come forth of the next door. I asked him how long he had lived there, finding that it was the place of his birth, I enquired of him, if he had ever heard anything of that Mr. Charnock. He told me he had heard his Mother (who died about 12 or 14 years since and was 80 years of age at her decease) often speake of him. That he kept a fire in, divers years; that his daughter lived with him, that once he was gon forth, and by her neglect (whome he trusted it with in his absence) the fire went out, and so all his work was lost. The Brasen head was very neere coming to speake, but soe was he disappointed. I suppose the pleasant-humoured man (for that he was soe appeares by his Breviary) [10] alluding to Friar Bacon's story [11] did so put off the inquisitiones of his simple neighbours, and thence it is come downe there by tradition till now.[12]

A copy of the manuscript roll alluded to still survives and is simply a transcription of some of the figures and tables from manuscripts of the Lullian treatises, including

[10] The title of Charnock's poem is "The breviary of alchemy."

[11] The legend that Roger Bacon made a head of brass that was capable of speech is very well known, and forms the subject of many stories.

[12] *Ambix*, II, p. 153.

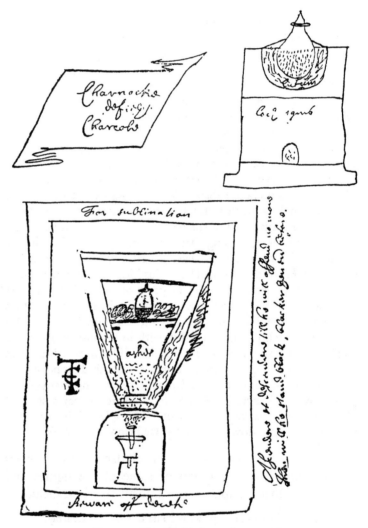

FIG. 17.—Charnock's lamp-furnace.

those of Plate I and Figure 13. Charnock's process consisted of what he called "circulations." Each circulation lasted a week; he performed at least 610 of those circulations, and had not even then obtained success.

These "circulations" were the main feature of the process based on the Lullian treatises and as practiced by the English alchemists. It is not possible to discover the exact practice, but we can at any rate discover something of what was intended.

"Circulation" in the Lullian works is a word that can be applied simply to the process we call "refluxing" today, the evaporation and condensation of a liquid (see p. 118), but which is also applied to the apparent successive conversion of the element earth into water, of water into air, of air into fire; and of fire back to earth again. This was accomplished by distillation either in stills or in vessels such as the pelican. Some solution placed in the vessel was, we may suppose, heated so as to be volatilized to "air," which condensed to "water" and perhaps an oily product, "fire," leaving behind a solid "earth" which could be redissolved in the liquid products and the process repeated indefinitely. There are indications that these circulations were intended to keep time with the circulation of the heavenly bodies. This would explain the period of a week, each day being assigned to one of the seven planets.

Neither Charnock nor Norton gives much information about the essentials of the alchemical process—what was to be used and what was to be done to it—though they both give much interesting detail about the incidental needs and troubles of the alchemist. Ripley, however, sets out the whole work very systematically in twelve stages, though

he speaks of each so obscurely that no one could discover
from his treatise what it was all about. Moreover, he is very
secret about his materials. But it is not difficult to see a general analogy between the type of process set out on pages
113-115 and Ripley's twelve stages. Thus we can tentatively
indicate the following stages of practice and theory.

1) *Calcination*

The reduction of the matters used to a non-metallic condition. This may correspond to the first reaction of the
acids with the gold and silver and distillation to dryness.

2) *Solution*

The action of the liquor distilled from the two solutions
upon the dry substances, bringing them into solution
again.

3) *Separation*

The further distillation of the liquid from the metallic
residues. This liquid has to be seven times distilled. (Cf.
p. 114)

4) *Conjunction*

The gold compound has three parts of its "water," the
silver compound nine parts; these are mixed, the vessel
closed and gently heated for many months.

5) *Putrefaction*

The matter becomes dark and thick, bubbles, settles,
and "putrefies." Shining colors like the rainbow develop.

6) *Congelation*

The colors disappear and the matter becomes dry and
white; this is the white stone and the end of the first stage.

7-12) *The Making of the Red Stone*

The second stage is designed to change the white stone
into the red one, and it is much more difficult to give it

any physical interpretation. It may be said that the process is a repetition of the stages gone before (except calcination, which is not required), ending with projection, the addition of the red stone to heated mercury, so transmuting it to gold.

This process is, I think, not susceptible of any physical explanation, because only the earlier stages corresponded to any physical changes. I think that we should not look for any explanation in terms of chemical composition, but rather for the significance that the things seen conveyed to the alchemist.

The alchemists undoubtedly performed a great many chemical operations on the matter, for it seems that each of these stages comprised a number of "circulations." In these processes, certainly performed with chemically impure materials, they must have seen a great many appearances that the chemist today would neglect. Thus the modern chemist when he distils a solution is not interested in the incidental appearances, whether the vapor is clear or cloudy, whether there is a scum on the surface, or in the exact details of the appearance of the sludge of solid residue. But these seem to have been the matters the alchemist was interested in; he concentrated attention on the form, color, and odor of the matter and watched intently all that occurred, relating it, not to chemical changes (a rational understanding of which was still far in the future), but to analogies drawn from the living world he observed about him and especially from the life of man.

The combination of two bodies he saw as *marriage*, the loss of their characteristic activity as *death*, the production of something new, as a *birth*, the rising up of vapors, as a *spirit leaving the corpse*, the formation of a volatile solid, as the making of a *spiritual body*. These conceptions influenced his idea of what should occur, and he therefore decided that the final end of the substances operated on should be analogous to the final end of man—a soul in a new, glorious body, with the qualities of clarity, subtlety, and agility.

We may, I think, regard the alchemy of this period as the practical pursuit and mental cultivation of the analogy between chemical changes and the life of man:

> Wherefore amonge Creatures theis two alone
> Be called *Microcosmus*, *Man* and our *Stone*.[13]

Some men pursued the renewal and glorification of matter, guiding themselves by this analogy, others the renewal and glorification of man, using the same analogy. Thus it is that we find alchemy to be at once a craft and a creed.

Since, moreover, the alchemical process was considered in terms of the phenomena of life, it could be best symbolized in these terms—which activity is the source of the enormously interesting series of alchemical pictures.

[13] Norton, "Ordinall," *Theatrum chemicum Britannicum*, p. 62, l.6.

XI

Alchemical Symbolism

THE use of symbolic pictures in alchemical texts goes back to the earliest times of alchemy, but it is not then highly developed. The figure of the serpent or dragon is the first symbol we meet with, and it represents matter in its imperfect unregenerate state. The dragon has to be slain, which means that the metals which are the subject of alchemy have to be reduced to a non-metallic condition and rendered susceptible of receiving a new spirit. Thus we remember that Zosimus wrote as early as the fourth century:

And that I may not write many things to you, my friend, build a temple of one stone, like ceruse in appearance, like alabaster, like marble of Proconnesus, having neither beginning nor end in its construction. Let it have within it a spring of pure water glittering like the sun. Notice on which side is the entry of the temple and, taking your sword in hand, so seek for the entry. For narrow is the place at which the temple opens. A serpent lies before the entry guarding the temple; seize him and sacrifice him. Skin him and, taking his flesh and bones, separate his parts; then reuniting the members with the bones at the entry of the temple, make of them a stepping stone, mount thereon, and enter. You will find there what you seek.[1]

[1] Berthelot, *Collection des anciens alchimistes grecs*, texte grec, p. 111.

The Alchemists

This symbol is found in alchemical works of every period.
Thus Charnock, more than a thousand years later, writes:

This is the philosophers dragon which eateth up his own Taile
Being famished in a doungell of glas and all for my prevail
Many yeres I kept this dragon in prison Strounge
Before I could mortiffy him, I thought it lounge
Yet at the lengthe by Gods grace yff ye beleve my worde
I vanquished him wythe a fyrie sword.

The dragon speaketh

——Souldiers in armour bright
Should not have kylled me in fyelde in fighte
Nor Mr. Charnock neither for all his philosophie
Yff by pryson and famyne he had not famysshed me.
Gye of Warwick nor Bevys of Southampton
Nere slew such a venemous dragon
Hercules fought with Hidra the Serpent
And yet he could not have his intent
Salamon the wyse, inclose too in a toome of brasse
But I was shutt upp in a doungeon off glass.
For my lyffe was so quick and my poyson soe stronge
That ere he could kill me it was full lounge.
Many [yeares] he hyld me in prison day and night
And kept me from sustenance to mynish me myght
But when I saw none other remedy
For very hunger I eate myne one bodye
And soe by corruption I became black and redd
But that precious stone that is in my hedd
Wyll be worth a Mille to him that hath skylle
And for that stone's sake he wysely dyd me kyll
My death I dyd him forgyve even at the very hower
Consydering that he wilbe beneficiall to the poore
For when I was alyve I was but strong poyson
And unprofitable for few things, in conclusion

146

To that I am now, dying in myne owne blood
For now I doe excell all other worldely good.
And a new name is given me of those that be wyse
For now I am named the Elixir of great price.
Which yff you will make prouffe, put to me my sister Mercury
and I will conjoyle (congeale) her into sylver in the twinkling
of an eye . . .[2]

FIG. 18.—The alchemical marriage, from the
Rosary of the Philosophers.

The dragon's sister is Mercury. As the dragon is matter,
metal, body, so his sister is spirit, metallic mercury, soul. We

[2] F. Sherwood Taylor, *op. cit.*, p. 154.

hear continually that "the dragon dieth not except with his sister," who is the agent in the alchemical process.

The second great symbol of alchemy is that of a marriage. The combination of Sol and Luna, "our gold" and "our silver," is symbolized in these terms, often with a frankness of sexual symbolism unacceptable in a modern published work. Sol is to impregnate Luna in order to generate the stone.

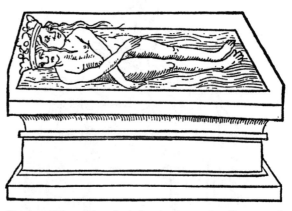

Fɪɢ. 19.—The alchemical death, from the *Rosary of the Philosophers.*

But in the Middle Ages the idea of impregnation and generation was very different from the present one, and it was symbolized as a death followed by a resurrection. Why is this? In any generation the form of the seed perishes, and a new being appears. In the generation of lower creatures, indeed, one being putrefies visibly and new creatures are generated seemingly without seed. So in all generation the feature that appealed intensely to the medieval mind is that expressed in the verses:

148

Amen, Amen, I say to you, unless the grain of wheat falling into the ground die, itself remaineth alone. But if it die it bringeth forth much fruit.

(John 12:24-25)

And likewise:

Senseless man, that which thou sowest is not quickened unless it die first. And that which thou sowest, thou sowest not the body that shall be: but bare grain, as of wheat, or of some of the rest. But God giveth it a body as he will: and to every seed a proper body.

(I Cor. 15:36-38) [3]

Thus the product of the marriage of Sol and Luna, figured as a hermaphrodite (or "rebis") because containing elements of both, is symbolized as a dead body, a hermaphroditic corpse in a tomb, becoming black and putrefied. As F. M. Van Helmont [4] said, the grave in the great world corresponds to the womb in the less world, a place of renewal, not of destruction. So, as the seed perishes, yet "God giveth it a body as he will." It is the "celestial virtue" that evokes the new form, the influence or spirit from above. The spirit of the dead rises up and the celestial influence descends. The simplest symbol that especially applies to this stage is the flying of the soul, shown as a small human figure, winged or otherwise, up to heaven. The celestial influence can be shown as dew descending, for dew was often identified with this heavenly influence. The other symbol is that of the birds who fly up to heaven and descend again: these are an obvious symbol for sublimation, distillation, and

[3] Quotations from the Bible are translations from the Vulgate, with which the alchemists were familiar.
[4] Cf. p. 175.

all the processes where a "spirit" is raised from a body. Similarly the *winged* figure of a hermaphrodite is the symbol of the spiritual body, the body wherein spirit has mastery over all the elements, the stone, white or red.

FIG. 20.—The rising up of the spirit, from the *Rosary of the Philosophers.*

Another symbol of the work is the tree, that which grows out of the earth, the mineral, and bears fruit, which is spiritual, having the power to become wine, which yields a spirit. Thus we see the trees bearing flasks or birds, also

150

PLATE IV. KHUNRATH'S "LABORATORIUM-ORATORIUM"

PLATE V. THE ALCHEMICAL PROCESS SHOWN PICTORIALLY [1-4]
(From the *Philosophia Reformata* of Mylius)

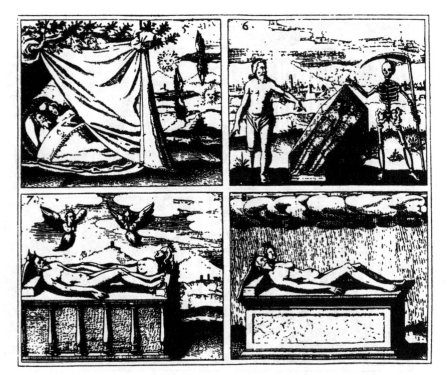

PLATE VI. THE ALCHEMICAL PROCESS SHOWN PICTORIALLY [5-8]
(From the *Philosophia Reformata* of Mylius)

PLATE VII. THE ALCHEMICAL PROCESS SHOWN PICTORIALLY [9-12]
(From the *Philosophia Reformata* of Mylius)

bearing fruits representing sun and moon, which stand as symbols for the red and the white stone.

An enormous number of analogies to various stages were evolved. The alchemical vessel could be thought of as an egg in which the matter was hatching; as the chamber in

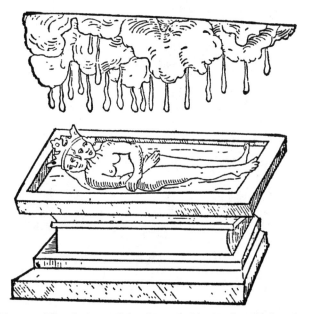

FIG. 21.—The descent of the dew, washing and purifying the corpse, from the *Rosary of the Philosophers*.

which was the bed of the pregnant mother who was to bring forth the child.

The devouring corrosive acid was a *lion*, the green lion, no doubt because of the color imparted to it by the copper compounds always present as impurity in the mixture of gold and silver.

Finally the work could be and was thought of in terms of religion, which was never absent from the medieval mind. Religion was familiar, the prime fact of life, and not some-

Fig. 22.—The winged hermaphrodite, symbolizing the red stone, from the *Rosary of the Philosophers.*

thing that was adverted to only on Sunday mornings. There was no thought of irreverence in applying it to everyday affairs. Thus the death of our Lord Jesus Christ and His resurrection in a glorified body was to the alchemist to be compared to the death of the metals and their rebirth as the

152

glorious stone. In the same way the Assumption of Our Lady, her rising up to heaven in the body, becoming a glorious body, there to be crowned by her Son, was also a type of the glorification of matter. The Trinity of three

Fig. 23.—The green lion, devouring the sun,
from the *Rosary of the Philosophers.*

Persons and one God was paralleled, in their minds, by the trinity of matter, e.g., salt, sulphur, and mercury in one body.

We often find the whole process set out in symbolic pictures. Sometimes these have a text accompanying them, sometimes they stand entirely by themselves. The five sets of four pictures comprising Plates V, VI, VII, VIII, and IX, taken from the *Philosophia Reformata* of John Daniel Mylius

FIG. 24.—The alchemical Resurrection, from the *Rosary of the Philosophers.*

(1622), are an example. It may be worth giving such notes upon them as the uninitiated author can furnish.

1. Gives a picture of the first matter of the work. The "two vapors" are seen at each side. The three statues give the animal, vegetable, and mineral "mercuries." Sun, moon, and *four* planets are shown; the fifth is Mercury, which is the subject of the picture.

154

2. Sol and Luna with the celestial Mercury above and the blood of the green lion, the solvent, issuing below.

3. and 4. Sol and Luna enter the bath where they are to be dissolved.

5. They are married: on the left the black birds attacking the sun and moon show the blackening and putrefaction of the bodies.

6. The bodies putrefying in the tomb of glass.

FIG. 25.—The alchemical Assumption, from the *Rosary of the Philosophers.*

7. Their souls depart: i.e., volatilization begins.
8. Sol and Luna become one hermaphrodite body which is watered by the dew of heaven, the celestial influence, identified with the condensed drops falling back.
9. The black crow appears, i.e., the blackened mass becomes fully volatile and the hermaphrodite body revives.
10. It is then the white stone having powers of healing and of transmutation into silver, typified by the tree of the moon with its silver fruit.

The second part of the process is more obscure, but is in outline a recapitulation of the first.

11. The white stone is used as a seed or ferment to initiate the process anew.
12. Gold is dissolved in the "mercury" with the addition of the white stone.
13. The body so formed is nourished with "mercury."
14. The bird descends into the body, meaning that the volatile part is fixed—made non-volatile.
15. The addition of more mercury and reiterated circulation increases the power and quantity of the medicine.
16. The stone is revived once more.
17. The red stone appears in its perfection. The tree of the sun is seen with its golden fruit. The serpent is in its power and also the green lion.

This is the end of the process. The remaining three pictures seem to be concerned with the whole work generally.

18. The green lion devouring the sun. The green lion is "our mercury," which is the solvent of gold.
19. Recapitulates the birth of the child, who is the stone, from Sol and Luna.
20. The King arises from his tomb, which figure may represent gold made by transmutation from base metal by aid of the stone.

These interpretations are, of course, tentative and are only intended to give a possible way of regarding such symbolic pictures. There are many such series employing many types of symbol. The *Twelve Keys* of Basil Valentine is another example, while the works of Michael Maier and of Stolcius contain an astonishing series of pictures which tell those who know how to look at them a great deal about the alchemical world view. But any attempt to interpret these would involve us in a mass of words and would indeed be useless, for these pictures are meant to be seen, not talked about. Our purpose is rather to understand the reason for this pictorial mode of expression.

In the first place, these alchemical symbols are not intended, as are the chemical symbols of today, as a brief means of expressing something concerning the bodies symbolized which could otherwise have been set out in words. It is true that Sol or the King can be confidently interpreted as gold, Luna as silver, and so forth, but the alchemical pictures meant very much more than that. They were, in fact, a way of understanding chemical change, of taking it into a mental scheme. The modern scientist can make sense of a chemical phenomenon by considering it in terms of the classes of bodies concerned and of their chemical composition: the phenomenon, e.g., of gold dissolving in *aqua regia* fits into the scheme of modern science and is at once linked a thousand ways to other phenomena having points of similarity; he can even picture it in terms of movements of particles.

Nothing of this was available to the alchemist, who had not conceived the idea of classifying chemical changes and had nothing that could, in our sense, be called a chemical science, into which they might be fitted. He had to explain

157

what he saw by finding analogies for it among his own ideas of the world. To the man of the Middle Ages, the important things in life were his relations with God and his neighbor—religion and human relationships—and the alchemical process became intelligible to him when expressed in those terms. To give to the combination of two substances to make a third the name or symbol of "a marriage and birth" was to fit the phenomenon into his world and so to make sense of it. He would then act on the principle that the phenomenon *was* a marriage and birth, and he would provide the sort of conditions which in his mind would be favorable to such a process.

Furthermore, the contemplation and practice of alchemy was not rewarded, as is modern science, by the intellectual satisfaction appropriate to one who finds the pieces of the puzzle fitting with admirable completeness, but by the emotional and spiritual satisfaction of one who sees living beings wonderfully fulfilling their ends appointed by God. The whole alchemical process has, as it were, a spiritual significance; it is a perfection of matter and was viewed with feelings appropriate to the sight of perfection. The alchemical process was a small illustration or example of the whole purpose of things, which were impelled to seek perfection by their striving towards the perfect ideas of their kind in God; it was likewise a symbol of Man whose end in life is to find bodily perfection in the glorious body, and spiritual fulfilment in the beatific vision of God.

So we may think that the alchemical picture was a truer expression of what alchemy was about than the alchemical book or recipe. The picture gave the inwardness of the process, expressing the meaning it had for the alchemist in terms that touched the deepest things in man; but it did not give any real chemical information. For that it was necessary to read the texts, comparing one with another, and above all to be instructed by a master.

Symbolic alchemy, in effect, presupposes that the changes in matter which it symbolizes are analogous to the changes in living beings and especially in man. It is, in fact, an understanding of nature in terms of life. And this is the reason why the symbols of alchemy are so widely applicable that certain authors have considered them to be merely a cover for the description of some mystical system by means of which, not the metals, but man was to be perfected.

It is quite true that some alchemical texts can be read as works of practical mysticism. Mrs. Atwood's *Suggestive Enquiry into the Hermetic Mystery* (1850) is a noble attempt to explain all alchemy in this way, but it must fail in view of the character of the texts, all the earlier of which betray evidence of unmistakable laboratory practice and knowledge of chemical technique. C. G. Jung,[5] in his recent *Psychologie und Alchimie* comes near to the truth when he claims that the alchemists, in studying matter symbolically, were also symbolizing their own mental content. This is true and is of interest to the psychologist and historian of religion; it is a theme which has recently been greatly developed. It is not, however, a complete picture of alchemy, because it concerns itself very little with what the alchemists were doing in their laboratories and what was the nature of the physical changes that they recorded in their writings and symbolized in their pictures. Alchemical pictures are good material for the psychologist, but to treat them as psychological material is to miss their real signifi-

[5] *Psychologie und Alchimie.* Zürich, 1944.

cance, which is the expression of the perfecting of matter in terms of human experience. If it is profitable to understand nature in human terms, alchemy has a present value; if not, it can interest us only as history.

By considering the symbolic representation of the alchemical process, however, we can understand how it was necessary for the development of alchemy to take either of two directions. The rise of natural science, beginning in a small way in the sixteenth century, vastly increasing after about 1650 and conquering all minds in the eighteenth century, made the parallel between the chemistry of the metals and the course of human relationships untenable. Furthermore, the publicity of the scientific method rendered the secret character of alchemy suspect. Consequently, alchemy could not maintain its medieval position. In some hands it tended to become chemistry, dropping its concentration upon gold and, more gradually, the analogies of chemical change with phenomena of life. Chemical preparations, the art of separating and combining bodies, is the part of alchemy that here survives. This process we see in the work of Paracelsus and his school of Libavius and many chemical writers of the seventeenth century.

In other hands alchemy tended to become more closely linked to religion and to provide a religious philosophy of nature and a mystical approach to her ways. Thus the "Hermetic Philosophy" was greatly cultivated in the seventeenth century by such men as Robert Fludd and Thomas Vaughan.[6] Their system was noble and impressive, but very slenderly linked to physical fact. It enjoyed a tremendous popularity in seventeenth-century Germany, but in England

[6] Cf. pp. 213-226.

seems to have been absorbed in the equally secret confraternity of the Masons.

These were the two chief changes in alchemy, in the direction of the practical and the mystical. Nevertheless, alchemy continued occasionally to be practiced by the old methods as far as they could be interpreted; but after about 1680 the men who practiced it were for the most part cranks or charlatans, and before 1850 the old tradition was dead.

Before we go on to consider these later developments of alchemy, it will be well to appraise some of the lively and appealing evidence for its truth which was offered to the public in the seventeenth century.

XII

Stories of Transmutations

IN the period after 1600, when skepticism concerning
the possibility of transmutation by alchemical means
was becoming commoner, there appeared a number of
circumstantial accounts of the process, which greatly forti-
fied the belief of those who were favorably inclined to the
Art. It is difficult to assess the evidential value of such accounts, especially since something unparalleled in the reader's experience is being recounted. If a modern president of
the Royal Society published an account of a transmutation
he had performed by means of a specimen of the stone, we
should question his sanity before we accepted his story.
How much less are we likely to credit stories some three
hundred years old! But even when we discount their age,
these stories are not extremely convincing. They are, however, very interesting, if only as showing what their writers
thought about transmutation. The three accounts printed in
this chapter are among the strongest, that is to say, they have
some small degree of confirmation through external events
or the credibility of the author.

Stories of Transmutations

The following account was first printed in French in 1612, and the English version which is here transcribed is a translation made in 1624 by someone who took the pseudonym of Eirenaeus Orandus. The spelling and punctuation are somewhat modernized.

Although that I *Nicholas Flammel, Notary,* and abiding in *Paris,* in this year one thousand three hundred fourscore and nineteen, and dwelling in my house in the street of Notaries, near unto the chapel of *St James* of the *Boucherie;* although, I say, that I learned but a little Latine, because of the small means of my parents, which nevertheless were by them that envie me most, accounted honest people; yet by the grace of God, and the intercession of the blessed Saints in *Paradise* of both sexes, and principally of Saint *James* of Gallicia, I have not wanted the understanding of the Books of the Philosophers, and in them learned their so hidden secrets. And for this cause there shall never be any moment of my life, when I remember this high good, wherein upon my knees (if the place will give me leave) or otherwise, in my heart with all my affection, I shall not render thanks to this most benign God, which never suffereth the child of the Just to beg from door to door, and deceiveth not them which wholly trust in his blessing.

Whilst therefore, I *Nicholas Flammel, Notary,* after the decease of my parents, got my living in our art of writing, by making inventories, dressing accounts, and summing up the expenses of tutors and pupils, there fell into my hands, for the sum of two florins, a gilded book, very old and large. It was not of paper or parchment, as other books be, but was only made of delicate rinds (as it seemed unto me) of tender young trees.[1] The cover of it was of brass, well bound, all engraven

[1] Probably a papyrus.

163

with letters or strange figures; and for my part, I think they might well be Greek characters, or some such like ancient language. Sure I am that I could not read them, and I know well they were not notes nor letters of the Latin, nor of the Gaul, for of them we understand a little. As for that which was within it, the leaves of bark or rind were engraven, and with admirable diligence written, with a point of iron, in fair and neat Latin letters colored.

It contained thrice seven leaves, for so they were counted in the top of the leaves, and always every seventh leaf was without any writing, but instead thereof, upon the first seventh leaf, there was painted a Virgin, and serpents swallowing her up; in the second seventh, a Cross whereon a serpent was crucified; and in the last seventh, there were painted deserts or wildernesses in the midst whereof ran many fair fountains, from whence there issued out a number of serpents, which ran up and down here and there. Upon the first of the leaves, was written in great capital letters of gold ABRAHAM THE JEW, PRINCE, PRIEST, LEVITE, ASTROLOGER, AND PHILOSOPHER, TO THE NATION OF THE JEWS, BY THE WRATH OF GOD DISPERSED AMONG THE GAULS, SENDETH HEALTH. After this it was filled with great execrations and curses (with this word MARANATHA, which was often repeated there) against every person that should cast his eyes upon it, if he were not Sacrificer or Scribe.

He that sold me this book, knew not what it was worth, no more than I when I bought it; I believe it had been stolen or taken from the miserable Jews; or found hid in some part of the ancient place of their abode. Within the book, in the second leaf, he comforted his Nation counselling them to fly vices, and above all idolatry, attending with sweet patience the coming of the Messias, which should vanquish all the Kings of the Earth, and should reign with his people in glory eternally. Without doubt this had been some very wise and understanding man. In the third leaf, and in all the other writings that followed,

164

to help his captive nation to pay their tributes unto the Roman emperors, and to do other things, which I will not speak of, he taught them in common words the transmutation of metals.

He painted the vessels by the sides and he advertised them of the colors, and of all the rest, saving of the *first agent*, of the which he spake not a word but only (as he said) in the fourth and fifth leaves entire he painted it, and figured it with very great cunning and workmanship, for although it was well and intelligibly figured and painted, yet no man could ever have beene able to understand it, without being well skilled in their Cabala, which goeth by tradition, and without having well studied their books. The fourth and fifth leaf therefore, was without any writing, all full of fair figures illuminated, or as it were illuminated, for the work was very exquisite.

First he painted a young man, with wings at his ankles, having in his hand a Caducean rod, writhen about with two Serpents, wherewith he struck upon a helmet which covered his head. He seemed to my small judgement to be the God Mercury of the Pagans. Against him there came running and flying with open wings a great old man, who upon his head had an hourglass fastened, and in his hands a hook (or scythe) like Death, with the which in terrible and furious manner, he would have cut off the feet of Mercury. On the other side of the fourth leaf, he painted a fair flower on the top of a very high mountain, which was sore shaken with the north wind: it had the foot blue, the flowers white and red, the leaves shining like fine gold, and round about it the dragons and griffons of the North made their nests and abode. On the fifth leaf there was a fair Rose tree flowered in the midst of a sweet garden, climbing up against a hollow oak, at the foot whereof boiled a fountain of most white water, which ran headlong down into the depths, notwithstanding it first passed among the hands of infinite people which digged in the earth seeking for it, but because they were blind, none of them knew it, except here and there one which considered the weight.

The Alchemists

On the last side of the fifth leaf, there was a King with a great falchion, who made to be killed in his presence by some soldiers a great multitude of little infants, whose mothers wept at the feet of the unpitiful soldiers; the blood of which infants was afterwards by other soldiers gathered up, and put in a great vessel, wherein the Sun and Moone came to bathe themselves. And because that this history did represent the more part of that of the Innocents slain by Herod, and that in this book I learned the greatest part of the Art, this was one of the Causes, why I placed in their Churchyard these Hieroglyphic symbols of this secret science. And thus you see that which was in the first five leaves.

I will not represent unto you that which was written in good and intelligible Latin in all the other written leaves, for God would punish me, because I should commit a greater wickedness than he who (as it is said) wished that all the men of the world had but one head that he might cut it off at one blow. Having with me therefore this fair book, I did nothing else day nor night, but study upon it, understanding very well all the operations that it shewed, but not knowing with what matter I should begin, which made me very heavy and solitary, and caused me to fetch many a sigh. My wife, Perrenelle, whom I loved as myself and had lately married, was much astonished at this, comforting me and earnestly demanding, if she could by any means deliver me from this trouble. I could not possibly hold my tongue but told her all, and showed her this fair book, whereof at the same instant that she saw it, she became, as much enamoured as myself, taking extreme pleasure to behold the fair cover, gravings, images and portraits, whereof, notwithstanding she understood as little as I, yet it was a great comfort to me to talk with her, and to entertain myself, what we should do to have the interpretation of them.

In the end I caused to be painted within my lodging, as naturally as I could, all the figures and portraits of the fourth and fifth leaf, which I showed to the greatest clerks in Paris, who

166

understood thereof no more than myself. I told them they were found in a book that taught the philosopher's stone, but the greatest part of them made a mock both of me, and of that blessed stone, excepting one called Master Anselm, which was a Licentiate in Physic, and studied hard in this science. He had a great desire to have seen my book, and there was nothing in the world which he would not have done for a sight of it, but I always told him that I had it not: only I made him a large description of the method.

He told me that the first portrait represented Time, which devoured all; and that according to the number of the six written leaves, there was required the space of six years, to perfect the stone; and then, he said, we must turn the glass and seethe it no more. And when I told him this was not painted, but only to show and teach the first agent (as was said in the book) he answered me, that this decoction for six years' space was, as it were, a second agent, and that certainly the first agent, which was there painted, was that white and heavy water, which without doubt was argent vive (quicksilver), which they could not fix, nor cut off his feet, that is to say, take away his volatility, save by long decoction in the purest blood of young infants, for in that, this quicksilver being joined with gold and silver, was first turned into a herb like that which was there painted, and afterwards by corruption, into serpents; which serpents being then wholly dried, and decocted by fire, were reduced into a powder of gold, which should be the stone.

This was the cause that during the space of one and twenty years, I tried a thousand broileries, yet never with blood, for that was wicked and villainous; for I found in my book that the philosophers called blood, the mineral spirit, which is in the metals, principally in the Sun, Moon, and Mercury, to the assembling whereof, I always tended; yet these interpretations for the most part were more subtle than true. Not seeing therefore in my works the signs, at the time written in my book, I was always to begin again.

The Alchemists

In the end having lost all hope of ever understanding those figures, for my last refuge, I made a vow to God, and *St. James* of *Gallicia*, to demand the interpretation of them at some Jewish priest, in some synagogue of Spain: wherupon, with the consent of Perrenelle, carrying with me the extract of the pictures, having taken the pilgrim's habit and staffe, in the same fashion as you may see me, without this same arch in the church-yard, in the which I put the hieroglyphical figures, where I have also set against the wall, on the one and the other side a procession, in which are represented by order all the colors of the stone, so as they come and go, with this writing in French:

> Moult plaist a Dieu procession
> S'elle est faicte en devotion:

that is,

> Much pleaseth God procession
> If't be done in devotion.

which is as it were the beginning of King Hercules his Book, which entreateth of the colors of the stone, entitled *Iris* or the *Rainbow*, in these terms, *Operis processio multum naturae placet*, that is *The procession of the work is very pleasant unto Nature*: the which I have put there expressly for the great Clerks who shall understand the allusion.

In this same fashion, I say, I put myselfe upon my way and so much I did, that I arrived at Montjoy, and afterwards at Saint James, where with great devotion I accomplished my vow. This done, in Leon at my return I met with a merchant of Boulogne, which made me known to a physician, a Jew by nation, and as then a Christian, dwelling in Leon aforesaid, who was very skilful in sublime sciences, called Master Canches.

As soon as I had shown him the figures of my extract, he being ravished with great astonishment and joy demanded of me incontinently if I could tell him any news of the book, from whence they were drawne? I answered him in Latin (wherein he asked me the question) that I hoped to have some good news of the book, if anybody could decipher unto me the enigmas. All at that instant transported with great ardor and joy, he began to decipher unto me the beginning. But to be short he [being] well content to learn news where this book should be, and I to hear him speak,—and certainly he had heard much discourse of the book, but (as he said) as of a thing which was believed to be utterly lost—, we resolved of our voyage, and from Leon we passed to Oviedo, and from thence to Sanson, where we put ourselves to Sea to come into France.

Our voyage had been fortunate enough, and already, since we were entered into this kingdom, he had most truly interpreted unto me the greatest parts of my figures, where, even unto the very points and pricks, he found great mysteries, which seemed unto me wonderful. When arriving at Orleans this learned man fell extremely sick, being afflicted with excessive vomitings, which remained still with him of those he had suffered at sea, and he was in such a continuall fear of my forsaking him, that he could imagine nothing like unto it. And although I was always by his side, yet would he incessantly call for me, but in sum he died, at the end of the seventh day of his sickness, by reason whereof I was much grieved, yet as well as I could, I caused him to be buried in the Church of the Holy Cross at Orleans, where he yet resteth; God have his soul, for he died a good Christian. And surely if I be not hindered by death I will give unto that Church some revenue, to cause some Masses to be said for his soul every day.

He that would see the manner of my arrival, and the joy of Perrenelle, let him look upon us two in this city of Paris, upon the door of the Chapel of St. James of the Boucherie, close by the one side of my house, where we are both painted, myself giving thanks at the feet of St. John, whom she had so often called upon. So it was, that by the grace of God and the intercession of the happy and holy Virgin and the blessed Saints James and John, I knew all that I desired, that is to say, the first

169

principles, yet not their first preparation, which is a thing most difficult, above all the things in the world. But in the end I had that also after long errors of three years, or thereabouts, during which time, I did nothing but study and labor, so as you may see me without this Arch, where I have placed my processions against the two pillars of it, under the feet of St. James and St. John, praying always to God, with my beads in my hand, reading attentively within a book, and weighing the words of the philosophers, and afterwards trying and proving the diverse operations which I imagined to myself, by their only words.

Finally I found that which I desired, which I also soon knew by the strong scent and odor thereof. Having this, I easily accomplished the mastery, for knowing the preparation of the first agents, and after following my book according to the letter I could not have missed it, though I would. Then the first time that I made projection was upon mercury whereof I turned half a pound, or thereabouts, unto pure silver, better than that of the mine, as I myself assayed, and made others assay many times. This was upon a Monday, the 17th of January, about noon, in my house, Perrenelle only being present, in the year of the restoring of mankind, 1382.

And afterwards, following always my book, from word to word, I made projection of the red stone upon the like quantity of mercury, in the presence likewise of Perrenelle only, in the same house the five and twentieth day of April following, the same year, about five o'clock in the evening, which I transmuted truely into almost as much pure gold, better assuredly than common gold, more soft and pliable. I may speak it with truth, I have made it three times, with the help of Perrenelle, who understood it as well as I because she helped me with my operations, and without doubt, if she would have enterprised to have done it alone, she had attained the end and perfection thereof. I had indeed enough when I had once done it, but I found exceeding great pleasure and delight in seeing and contemplating the admirable works of nature with the vessels.

To signify unto thee then, how I have done it three times, thou shalt see in this Arch, if thou have any skill to know them, three furnaces, like unto them which serve for our operations. I was afraid for a long time, that Perrenelle could not hide the extreme joy of her felicity, which I measured by my own, and lest she should let fall some word amongst her kindred, of the great treasures which we possessed, for extreme joy takes away the understanding as well as great heaviness, but the goodness of the most great God had not only filled me with this blessing, to give me a wife chaste and sage (for she was moreover not only capable of reason, but also to do all that was reasonable), and more discreet and secret than ordinarily other women are. Above all, she was exceedingly devout, and therefore seeing her self without hope of children, and now well stricken in years, she began, as I did, to think of God and to give ourselves to the works of mercy.

At that time when I wrote this Commentary in the year one thousand four hundred and thirteen, in the end of the year, after the decease of my faithful companion which I shall lament all the days of my life, she and I had already founded, and endowed with revenues 14 hospitals in this City of Paris, we had new built from the ground three chapels, we had enriched with great gifts and good rents, seven churches, with many reparations in their churchyards, besides that which we have done at Boulogne, which is not much less than we have done here. I will not speak of the good which both of us have done to particular poor folks, principally to widows and poor orphans, whose names if I should tell and how I did it, besides that my reward should be given me in this world, I should likewise do displeasure to those good persons, whom I pray God blesse, which I would not do for anything in the world.

Building therefore these churches, churchyards, and hospitals in this City, I resolved myself to cause to be painted in the fourth Arch of the Churchyard of the Innocents, as you enter in by the great gate in St. Dennis Street and taking the way on the right hand, the most true and essential marks of the Art, yet under veils and hieroglyphical covertures, in imitation of those which are in the gilded book of Abraham the Jew, which may represent two things, according to the understanding and capacity of them that behold them.

First, the mysteries of our future and undoubted Resurrec-

Fig. 26.—The Archway, painted by Flamel, as illustrated in the edition of 1612.

tion, at the day of Judgement and coming of good Jesus (whom may it please to have mercy upon us) a history which is well agreeing to a churchyard. And secondly they may signify to them which are skilled in Natural Philosophy, all the principal and necessary operations of the mastery. These hieroglyphic figures shall serve as two ways to lead into the heavenly life; the first and most open sense teaching the sacred mysteries of

172

our salvation; (as I will show hereafter) the other teaching every man that hath any small understanding in the stone, the linear way of the work, which being perfected by any one, the change of evil into good, takes away from him the root of all sin (which is covetousness) making him likeable, gentle, pious, religious, and fearing God, how evil soever he was before. For from thenceforward he is continually ravished with the great grace and mercy which he hath obtained from God, and with the profoundness of his Divine and admirable works.

These are the reasons that have moved me to set these forms in this fashion, and in this place, which is a churchyard, to the end that if any man obtains this inestimable good to conquer this rich golden fleece, he may think with himself (as I did) not to keep the talent of God digged in the earth, buying lands and possessions which are the vanities of this world, but rather to work charitably towards his brethren, remembering himself that he learned this secret among the bones of the dead, in whose number he shall shortly be found and that after this life he must render an account, before a just and redoubtable Judge, which will censure even to an idle and vain word.

Let him therefore, which having well weighed my words, and well known and understood my figures, hath first gotten elsewhere the knowledge of the first beginnings and agents (for certainly in these figures and commentaries, he shall not find any step or information thereof) perfect to the glory of God the mastery of Hermes, remembering himself of the Church Catholic Apostolic and Roman; and of all other churches, churchyards and hospitals, and above all the Church of the Innocents in this City (in the churchyard whereof he shall have contemplated these true demonstrations), opening bounteously his purse to them that are secretly poor honest people, desolate weak women, widows and forlorn orphans. So be it.[2]

[2] *Nicholas Flammel, his exposition of the hieroglyphicall figures,* which he caused to be painted upon an arch in St. Innocent's churchyard in Paris, by Eirenaeus Orandus. London, 1624.

Much of this remarkable story can, of course, receive no confirmation, but it seems quite certain that there was a Nicholas Flamel who lived in the house mentioned in the story, gave a great deal of money to charitable objects, and was interested in alchemy. The figures in the archway, as depicted in Figure 26, survived from 1407 until the eighteenth century and a marble tablet from his tomb, now in the Musée de Cluny,[3] records that Nicolas Flamel, formerly a scrivener, left to the church (of St.-Jacques-la-Boucherie) certain rents and houses that he had bought in his lifetime and had made gifts to various churches and hospitals in Paris. The tomb is carved in low relief with figures of Christ, of St. Peter and St. Paul, and between these figures representations of the sun and moon, which, with the inscriptions on the archway, attest his connection with alchemy.

There has been a good deal of theorizing by disbelievers in alchemy about the manner in which Flamel made the large fortune to which his gifts bear witness, but it is not surprising that at this distance of time nothing very convincing can be said on the subject. It is certain, however, that the history of Flamel was a great source of belief in alchemy, both in the fifteenth century and again in the seventeenth, after the narrative had been printed.

2. THE TESTIMONY OF VAN HELMONT

Jean Baptiste van Helmont was born in 1577. He was widely read in all the sciences, especially chemistry, physiology, and medicine, which he practiced from 1599 to his death in 1644. He was a disciple of Paracelsus and a believer

[3] Musée des Thermes et de l'Hôtel de Cluny. *Catalogue général.* Paris, 1922, Vol. I, p. 105, No. 574.

in the spiritual view of nature; but none the less he made important chemical discoveries. He was the first to realize that there were gases other than air, and the very word "gas" was his invention. He proved experimentally that only a very small proportion of a growing plant was derived from the earth, by growing a willow slip in a weighted quantity of earth and showing that the diminution of the weight of the earth was negligible compared with the increase of weight of the willow. This was a fine piece of experimental work, which is not diminished by the fact that he supposed the substance of the plant to be almost entirely formed from water by a sort of transmutation, not realizing the part played by the air in its formation.

His works were collected and published in 1648 by his son, Francis Mercurius van Helmont, who was much less scientific and more superstitious than his father. We, therefore, cannot be sure that J. B. van Helmont wrote this account of transmutation as we have it, but it cannot be said that it is not in character with the greatest part of his work. Furthermore, Francis Mercurius van Helmont declares his disbelief in alchemy in the *Paradoxal Discourses concerning the Macrocosm and Microcosm* (1685), and if this was his opinion in 1648, he is unlikely to have invented the story which follows:

For truly, I have divers times seen it and handled it with my hands, but it was of color, such as is saffron in its powder, yet weighty and shining like unto powdered glass. There was once given unto me one fourth part of one grain (but I call a grain the six hundredth part of one ounce): this quarter of one grain therefore, being rolled up in paper, I projected upon eight ounces of quicksilver made hot in a crucible; and straightway all the quicksilver, with a certain degree of noise, stood still from flowing, and being congealed, settled like unto a yellow lump: but after pouring it out, the bellows blowing, there were found eight ounces and a little less than eleven grains of the purest gold. Therefore one only gram of that powder had transchanged 19186 parts of quicksilver, equal to itself, into the best gold.[4]

The account has internal evidence against it. Since the density of gold is 19.3 and that of mercury is 13.6, the sudden conversion of the latter into the former should be accompanied by a shrinkage of about one third, which would be very noticeable but is not mentioned here.

But the most interesting and circumstantial entry is concerned with the use of the stone as a medicine:

There was a certain Irishman, whose name was Butler, being sometime great with James King of England. He being detained in the prison of the castle of Vilvord, and taking pity on Baillius a certain Franciscan Monk, a most famous preacher of Gallo-Brittain, who was also imprisoned, having a formidable Erysipelas in his arm, on a certain evening when as the sick monk did almost despair, he swiftly tinged a certain little stone in a spoonful of almond milk, and presently withdrew it thence. But he said unto the keeper of the prison, reach this supping to that monk, and how much soever he shall take thereupon, he shall be whole at least within a short hours space; which thing came to pass with the greatest admiration of the keeper, and the sick man not knowing from whence so sudden health had shined on him, seeing he was ignorant that he had taken anything. For his left arm, being before hugely swollen, fell down as that it could presently scarce be discerned from the other. On the morning following, I being entreated by great men, came to Vilvord as a witness of his deeds: therefore I contracted a friendship with Butler.

[4] J. B. van Helmont, *Oriatrike or physick refined*. London, 1662, pp. 751-752.

·Presently afterwards, I saw a poor old woman, a laundress, who from 16 years or thereabouts laboured with an intolerable megrim,[5] presently cured in my presence. Indeed he by the way, or lightly, dipt the same little stone in a spoonful of oil of olives, and presently cleansed the little stone by licking of it, and laid it up into the sheath of his breast, but that spoonful of oil, he poured into a small bottle of oil, whereof one only drop he commanded to be anointed on the head of the aforesaid old woman, who was thereby straightway cured and remained whole for some years, the which I attest. I was amazed, as if he were become another Midas, but he smiling on me said:

"My most dear Friend unless thou come thitherto, so as to be able by one only remedy, to cure every disease, thou shalt remain in thy young beginnings, however old thou shalt become."

I easily assented thereto, because I had learned that thing from the Secrets of Paracelsus, and being now more confirmed by sight and hope. But I confess with a willing mind that that new manner of curing was unaccustomed and unknown to me. I therefore said, that a young prince of our Court, Viscount of Gaunt, brother to the Prince of Epifuoy, of a very great house, was so wholly prostrated by the Gout, that he henceforth lay only on one side, being wretched, and deformed with many knots, he therefore taking hold of my right hand, said "Wilt thou that I cure that young man; I will cure him for thy sake." But I replied: "But he is of that obstinacy, that he had rather die, than to drink even but one only medicinal potion." "Be it so," said Butler, "For neither do I require any other thing, than that he do every morning touch the little Stone which thou seest, with the tip of his tongue. For after three weeks from thence, let him wash the painful and unpainful knots daily with his own urine, and thou shalt soon afterwards see him cured, and soundly walking go thy way, and tell him with joy what I have said."

[5] Migraine, sick headache.

I therefore being glad, returned to Brussels, and tell him what Butler had said.

But the Potentate answered "Go to tell Butler, that if he restore me, as thou hast said, I will give him as much as he shall require: demand the price and I will willingly sequester that which is deposited, for his security." And when I declared that thing to Butler on the day following, he was wroth and said "That Prince is mad, or witless, and miserable, and therefore neither will I ever help him, for neither do I stand in need of his money, neither do I yield or am I inferior unto him." Yea, neither could I ever induce him to perform what he had before promised, wherefore I began to doubt lest the foregoing things which I had seen, were as it were dreams.

It happened in the meantime that a friend, overseer and master of the glass furnace at Antwerp, being exceeding fat, most earnestly requested of Butler to be freed from the trouble of his fatness, unto whom Butler offered a small piece of that little stone, that he might once every morning lick or speedily touch it with the top of his tongue. And within three weeks I saw his breast made more strait or narrow by one span, and him to have lived no less whole afterward. Wherefore I began again to believe that the same thing might have happened in the aforesaid gouty Prince, which he had promised.

In the meantime, I sent to Vilvord, to Butler for a remedy, in the case of poison occasionally given me by a secret enemy. For I miserably languished, all my joints were pained and my pulse vehement, being at length become an intermitting one, did accompany the faintings of my mind, and extinguishment of my strength. Butler being as yet detained in prison, forthwith commanded my household servant whom I had sent that he should bring unto him a small bottle of oil of olives, and his little stone aforesaid being tinged therein (as at other times) he sent that oil unto me; and bade him that with one only small drop, I should anoint only one place of the pain or all particular places if I would, the which I did, and yet felt no help thereby . . .

My wife was now for some months oppressed with a pain of the muscle of her right arm, so as that she could neither lift up her hand, and much less lift anything upwards. And moreover by reason of grief and sorrow for me, she now by degrees languished in both her legs, from the foot even to the groin, with a cruel œdema, the which did in its pit, show the trace of one's finger dipped into it even unto the second joint. For because she had contracted these œdemas by reason of the grief for my tribulation, a medicine was despised so long as her grief ceased not. She, therefore, seeing the work of Butler's oil to be vain on me, and being willing before some gentlewomen to mock my credulity, anointed one only drop of that oil on her right arm, and straightway, it being freely moved was beyond hope restored, together with its former strength: we all admired at the wonder of so sudden an event: wherefore she anointed the ankles of both her legs with one only drop on both sides, being spread about on the circle of the ankle, and presently within less than a quarter of an hour, all the œdema vanishes away: she also, through God's favour, liveth as yet nineteen years since, in health.[6]

3. THE TESTIMONY OF HELVETIUS

Johann Friedrich Schweitzer, who, as the custom then was, latinized his name to Helvetius, was born at Köthen in the Duchy of Anhalt. He was a well-known medical man, author of one or two medical and botanical books, and physician to the Prince of Orange. There seems to be no doubt that he was the author of the work entitled *The Golden Calf*, from an abbreviated translation of which the following account of a transmutation was taken. There

[6] Van Helmont, *op. cit.*, pp. 587-589.

seems in this account to be no room for any mistake or illusion: Helvetius either transmuted lead to gold or has lied prodigiously.

The twenty seventh of *December,* 1666, in the afternoon, came a Stranger to my house at the Hague, in a plebeian habit, honest Gravity, and serious authority; of a mean Stature, a little long face, with a few small pock holes, and most black hair, not at all curled, a beardless chin, about three or four and forty years of age (as I guessed) and born in *North Holland.* After salutation he beseeched me with a great reverence to pardon his rude accesses, being a great lover of the Pyrotechnyan Art; adding, he formerly endeavored to visit me with a friend of his, and told me he had read some of my small treatises; and particularly, that against the Sympathetick Powder of Sir *Kenelm Digby,* and observed my doubtfulness of the philosophical mystery, which caused him to take this opportunity, and asked me if I could not believe such a medicine was in nature, which could cure all diseases, unless the principal parts (as lungs, liver &c.) were perished, or the predestinated time of death were come.

To which I replied, I never met with an Adept, or saw such a medicine, though I read much of it, and have wished for it. Then I asked if he were a physician, but he preventing my question, said he was a founder of brass, yet from his youth learnt many rare things in chymistry, of a friend particularly, the manner to extract out of metals many medicinal arcanas by force of fire, and was still a lover of it.

After other large discourse of experiments in metals this *Elias* asked me if I could know the philosopher's stone when I see it, I answered not at all, though I had read much of it in *Paracelsus, Helmont, Basilius,* and others; yet dare I not say I could know the philosopher's matter. In the interim he took out of his bosom pouch or pocket, a neat ivory box, and out of it took three ponderous pieces or small lumps of the stone, each about the bigness of a small walnut, transparent, of a pale brimstone color, whereunto did stick the internal scales of the crucible, wherein it appeared this most noble substance was melted; the value of them might be judged worth about twenty tons of gold, which when I had greedily seen and handled almost a quarter of an hour, and drawn from the owner many rare secrets of its admirable effects in human and metallic bodies, and other magical properties, I returned him this treasure of treasures; truly with a most sorrowful mind, after the custom of those who conquer themselves, yet (as was but just) very thankfully and humbly, I further desired to know why the color was yellow, and not red, ruby color, or purple, as philosophers write; he answered, that was nothing, for the matter was mature and ripe enough.

Then I humbly requested him to bestow a little piece of the medicine on me, in perpetual memory of him, though but the quantity of a coriander or hemp seed. He presently answered. Oh no, no, this is not lawful though thou wouldst give me as many ducats in gold as would fill this room, not for the value of the matter, but for some particular consequences, nay, if it were possible (said he) that fire could be burnt of fire, I would rather at this instant cast all this substance into the fiercest flames.

But after he demanding, if I had another private chamber, whose prospect was from the public street, I presently conducted him in to the best furnished room backwards, where he entred without wiping his Shoes (full of snow and dirt) according to the custom in *Holland*, then not doubting but he would bestow part thereof, or some great secret treasure on me, but in vain; for he asked for a little piece of gold and pulling off his cloak or pastoral habit, opened his doublet, under which he wore five pieces of gold hanging in green silk ribons, as large as the inward round of a small pewter trencher: and this gold so far excelled mine, that there was no comparison, for flexibility and color; and these figures with the inscriptions

181

ingraven, were the resemblance of them, which he granted me to write out. [*The author here illustrates these medals.*]

I being herewith affected with great admiration, desired to know where and how he came by them. Who answered, An outlandish friend who dwelt some days in my house (giving out he was a lover of this art, and came to reveal this art to me) taught me various arts: First, How out of ordinary stones and christalls, to make rubies, chrysolites and sapphires, &c. much fairer then the ordinary. And how in a quarter of an hour to make crocus martis, of which one dose would infallibly cure the pestilential dysentery (or bloody flux) and how to make a metallic liquor most certainly to cure all kinds of dropsies in four days; as also a limpid clear water sweeter than honey, by which in two hours of itself, in hot sand, it would extract the tincture of *granats, corals, glasses,* and such like more, which I *Helvetius* did not observe.

My mind being drawn beyond those bounds, to understand how such a noble juice might be drawn out of the metals, to transmute metals; but the shade in the water deceived the dog of the morsel of flesh in his mouth. Moreover he told me his said master caused him to bring a glass full of rain water, and fetch some refined silver laminated in thin plates, which therein was dissolved within a quarter of an hour, like ice when heated: And presently he drank to me the half, and I pledged him the other half, which had not so much taste as sweet milk; whereby me thought I became very light hearted. I thereupon asked if this were a philosophical drink, and wherefore we drank this potion? He replied I ought not to be so curious. And after he told me that by the said masters directions, he took a piece of a leaden pipe, gutter or cistern, and being melted put a little such sulphurious powder out of his Pocket and once again put a little more on the point of a knife, and after a great blast of bellows in short time poured it on the red stones of the kitchen chimney, which proved most excellent pure Gold; which he said brought him into such a trembling amazement, that he

PLATE VIII. THE ALCHEMICAL PROCESS SHOWN PICTORIALLY [13-16]
(From the *Philosophia Reformata* of Mylius)

PLATE IX. THE ALCHEMICAL PROCESS SHOWN PICTORIALLY [17-20]
(From the *Philosophia Reformata* of Mylius)

PLATE X. MERCURY
(From Thurneysser's *Quinta Essentia*, 1570)

PLATE XI. SULPHUR
(From Thurneysser's *Quinta Essentia*, 1570)

could hardly speak. But his master thereupon again encouraged him, saying, cut for thy self the sixteenth part of this for a memorial, and the rest give away amongst the poor which he did. And he distributed so great an alms as he affirmed (if my memory fail not) to the Church of *Sparrenda*: but whether he gave it at several times or once, or in the golden masse, or in silver coin, I did not ask.

At last said he (going on with the story of his master) he taught me thoroughly this almost divine art. As soon as his history was finished, I most humbly begged he would shew me the effect of transmutation to confirm my faith therein, but he dismissed me for that time in such a discreet manner, that I had a denial. But withall promising to come again at three weeks end, and shew me some curious arts in the fire, and the manner of projection, provided it were then lawful without prohibition.

And at the three weeks end he came, and invited me abroad for an hour or two, and in our walks having discourses of divers of natures secrets in the fire; but he was very sparing of the great elixir, gravely asserting, that was only to magnify the most sweet fame, and name of the most glorious God; and that few men endeavored to sacrifice to him in good works, and this he expressed as a pastor or minister of a church; but now and then I kept his ears open, intreating to shew me the metallic transmutation; desiring also he would think me so worthy to eat and drink and lodge at my house, which I did prosecute so eagerly, that scarce any suitor could plead more to obtain his mistress from his corrival; but he was of so fixt and stedfast a spirit, that all my endeavors were frustrate: yet I could not forbear to tell him further I had a fit laboratory, and things ready and fit for an experiment, and that a promised favour was a kind of debt; yea, true said he, but I promised to teach thee at my return with this proviso, if it were not forbidden.

When I perceived all this in vain, I earnestly craved but a

most small crumb or parcel of his powder or stone, to transmute four grains of lead to gold; and at last out of his philosophical commiseration, he gave me a crumb as big as a rape or
turnip feed; saying, receive this small parcel of the greatest
treasure of the world, which truly few kings or princes have
ever known or seen: But I said, This perhaps will not transmute
four grains of lead, whereupon he bid me deliver it him back,
which in hopes of a greater parcel I did; but he cutting halfe off
with his nail, flung it into the fire, and gave me the rest wrapped
neatly up in blue paper; saying, It is yet sufficient for thee.
I answered him (indeed with a most dejected countenance)
Sir, what means this? the other being too little, you give me
now less.

He told me, If thou canst not manage this; yet for its great
proportion for so small a quantity of lead, then put into the
crucible two drams, or half an ounce, or a little more of the
lead; for there ought no more lead be put in the crucible than
the medicine can work upon, and transmute. So I gave him
great thanks for my diminished Treasure, concentrated truly
in the superlative degree, and put the same charily up into my
little box; saying I meant to try it the next day; nor would I
reveal it to any.

Not so, not so; (said he) for we ought to divulge all things
to the children of art; which may tend to the singular honor of
God, that so thay may live in the theosophical truth, and not
at all die sophistically. After I made my confession to him, that
whilst this mass of his medicine was in my hands, I endeavored
to scrape a little of it away with my nail, and could not forbear;
but scratcht off nothing, or so very little, that it was but as
an indivisible atom, which being purged from my nail, and
wrapt in a paper; I projected on lead, but found no transmutation; but almost the whole masse of lead flew away, and the
remainder turned into a meer glassy earth; at which unexpected
passage, he smiling, said, thou art more dexterous to commit
theft, then to apply thy medicine; for if thou hadst only wraped

up thy stolen prey in yellow wax, to preserve it from the arising fumes of lead, it would have penetrated to the bottom of
the lead, and transmuted it to gold; but having cast it into the
fumes, partly by violence of the vaporous fumes, and partly by
the sympathetic alliance, it carryed thy medicine quite away:
For gold, silver, quick-silver, and the like metals, are corrupted
and turn brittle like to glass, by the vapors of lead.

Whereupon I brought him my crucible wherein it was done,
and instantly he perceived a most beautiful saffron-like tincture
stick on the sides; and promised to come next morning, by
nine in the morning, and then would shew me my error, and
the said medicine should transmute the lead into gold. Nevertheless I earnestly prayed him in the interim to be pleased to
declare only for my present instruction, if the philosophic
work cost much, or required long time.

My friend, my friend (said he), thou art too curious to know
all things in an instant, yet will I discover so much; that neither
the great charge, or length of time, can discourage any; for as
for the matter, out of which our magistery is made, I would
have thee know there is only two metals and minerals, out of
which it is prepared; but in regard the sulphur of philosophers
is much more plentiful and abundant in the minerals; therefore
it is made out of the minerals. Then I asked again, What was
the menstruum, and whether the operation or working were
done in glasses, or crucibles?

He answered, the Menstruum was a heavenly salt, or of a
heavenly virtue, by whose benefit only the wise men dissolve
the earthly metallic body, and by such a solution is easily and
instantly brought forth the most noble elixir of philosophers.
But in a crucible is all the operation done and performed, from
the beginning to the very end, in a open fire, and all the whole
work is no longer from the very first to the last then four days,
and the whole work no more charge than three florens; and
further, neither the mineral, out of which, nor the salt, by
which it was performed, was of any great price. And when I

replied, the philosophers affirm in their writings, that seven or nine months at the least, are required for this work.

He answered, Their writings are only to be understood by the true adeptists; wherefore concerning time they would write nothing certain: Nay, without the communication of a true adept philosopher, not one student can find the way to prepare this great magistery, for which cause I warn and charge thee (as a friend) not to fling away thy money and goods to hunt out this art; for thou shalt never find it. To which I replied thy master, (though unknown) shewed it thee; So mayst thou perchance discover something to me, that having overcome the rudiments, I may find the rest with little difficulty, according to the old saying. *It is easier to add to a foundation, than begin a new.*

He answered, In this art 'tis quite otherwise; for unless thou knowest the thing from the head to the heel, from the eggs to the apples; that, from the very beginning to the very end thou knowest nothing, and though I have told thee enough; yet thou knowest not how the philosophers do make, and break open the glassy seal of Hermes, in which the Sun sends forth a great splendor with his marvellous colored metallic rayes, and in which looking glass the eyes of Narcissus behold the transmutable metals, for out of those ways the true adept philosophers gather their fire; by whose help the volatile metals may be fixed into the most permanent metals, either gold or silver.

But enough at present; for I intend (God willing) once more to morrow at the ninth hour (as I said) to meet, and discourse further on this philosophical subject, and shall shew you the manner of projection. And having taken his leave, he left me sorrowfully expecting him; but the next day he came not, nor ever since: Only he sent an excuse at half an hour past nine that morning, by reason of his great business, and promised to come at three in the afternoon, but never came, nor have I heard of him since; whereupon I began to doubt of the whole matter.

Stories of Transmutations

Nevertheless late that night my wife (who was a most curious student and enquirer after the art, whereof that worthy man had discourst) came soliciting and vexing me to make experiment of that little spark of his bounty in that art, whereby to be the more assured of the truth; saying to me, unless this be done, I shall have no rest nor sleep all this night, but I wished her to have patience till next morning to expect this Elias; saying, perhaps he will return again to shew us the right manner.

In the meantime (she being so earnest) I commanded a fire to be made (thinking alas) now is this man (though so divine in discourse) found guilty of falsehood; and secondly attributing the error of my projecting the grand theft of his powder in the dirt of my nail to his charge, because it transmuted not the lead that time; and lastly, because he gave me too small a proportion of his said medicine (as I thought) to work upon so great a quantity of lead as he pretended and appointed for it. Saying further to myself, I fear, I fear indeed this man hath deluded me; Nevertheless my wife wrapped the said matter in wax, and I cut half an ounce or six drams of old lead, and put into a crucible in the fire, which being melted, my wife put in the said Medicine made up into a small pill or button, which presently made such a hissing and bubbling in its perfect operation, that within a quarter of an hour all the mass of lead was totally transmuted into the best and finest gold, which made us all amazed as planet-struck.

And indeed (had I lived in *Ovids* Age) there could not have been a rarer metamorphosis than this, by the art of alchemy. Yea, could I have enjoyed Argus's eyes, with a hundred more, I could not sufficiently gaze upon this so admirable and almost miraculous a work of nature; for this melted lead (after projection) shewed us on the fire the rarest and most beautiful colors imaginable; yea, and the greenest color, which as soon as I poured forth into an ingot, it got the lively fresh color of blood; and being cold shined as the purest and most refined transplendent gold. Truly I, and all standing about me, were exceedingly startled, and did run with this aurified lead (being yet hot) unto the goldsmith, who wondered at the fineness, and after a short trial of touch, he judg'd it the most excellent gold in the whole world, and offered to give most willingly fifty florins for every ounce of it.

The next day a rumor went about the *Hague*, and spread abroad; so that many illustrious persons and students gave me their friendly visits for its sake: Amongst the rest the general Assay-Master, or Examiner of the Coins of this Province of *Holland*, Mr. Porelius, who with others earnestly beseeched me to pass some part of it through all their customary trials, which I did, the rather to gratifie my own curiosity.

Thereupon we went to Mr. *Buectel* a silversmith, who first tried *per quartam*, viz. he mixt three or four parts of silver with one part of the said gold, and laminated, filed or granulated it, and put a sufficient quantity of *Aqua Fort* thereto, which presently dissolved the silver, and suffered the said gold to precipitate to the bottom; which being decanted off, and the calx or powder of gold dulcified with water, and then reduced and melted into a body, became excellent gold: And whereas we feared loss, we found that each dram of the said first gold was yet increased, and had transmuted a scruple of the said silver into gold, by reason of its great and excellent abounding tincture.

But now doubting further whether the silver was sufficiently separated from the said gold, we instantly mingled it with seven parts of antimony, which we melted and poured into a cone, & blowed off the regulus on a test, where we missed eight grains of our gold, but after we blowed away the rest of the antimony, or superfluous scoria, we found nine grains of gold more for our eight grains missing, yet this was somewhat pale and silver-like, which easily recovered its full color afterwards, So that in the best proof of fire we lost nothing at all of this gold; but gained as aforesaid. The which proof again I repeated thrice, and found it still alike, and the said remaining silver out

188

of the aqua fortis, was of the very best flexible silver that could be; so that in the total, the said medicine (or elixir) had transmuted six drams and two scruples of the lead and silver, into most pure gold.[7]

These stories serve at least to illustrate that in the seventeenth century there was written evidence of transmutation strong enough to confirm the belief of those inclined to alchemy. Doubtless it is now forever impossible to resolve the question whether the authors of these stories really saw something that appeared to be a transmutation, whether they told of themselves what they had heard from others, or whether they resembled

> . . . Sir Agrippa, for profound
> And solid lying much renowned.

The stories remain as curiosities, the more fascinating because inexplicable.

There were, indeed, and perhaps are, specimens of alchemically prepared gold to be seen. Thus John Evelyn in 1644 visited Florence and saw in a museum "an yron nayle, one halfe whereof, being converted into gold by one Thornheuser, a German chymist, is look'd on as a great rarity, but it plainly appeared to have been soldered together." There exist also medals purporting to be struck from alchemical gold, but, alas, one at least of these is only gilt!

[7] Cooper, William, bookseller, *The philosophical epitaph* of W. C(ooper) . . . also, *a brief of the Golden Calf* . . . by J. F. Helvetius . . . London, 1673-1675.

XIII

From Alchemy to Chemistry

I'T is often taken for granted that alchemy at some stage of its career developed into chemistry. This is very much too simple an account of what occurred in the two centuries between the time of, let us say, Paracelsus and Boerhaave. The truth is that one part of alchemy became with little change one part of chemistry, and the part that was so transferred was its laboratory technique. But it cannot be too much emphasized that the intention of seeking knowledge of particular things which is vital to chemistry, the scientific method which made it a science, and the atomic speculations that characterized its explanations, came from other sources.

Thus alchemy is distinguished from chemistry first by its purpose and secondly by its method. The purpose of alchemy is the perfection of all things in their kind and most especially of metals; that of chemistry is the gaining of knowledge concerning different kinds of matter and the use of this knowledge for all manner of ends.

The method of alchemy is primarily the study of alchemical texts whose authors were presumed to have known the process that their readers sought to discover; secondarily it is a reflection upon nature, as known by common sense, in order to discover her general laws and apply these to the alchemical problem; thirdly and very much lastly, it is experiment, which in any case was limited to endeavors to find out the conditions in which the appearances prescribed by the texts could be attained. The method of chemistry is the accurate description of changes of all kinds of matter and the classification of such changes in order to discover general laws. To the chemist, books are storehouses, not authorities; analogies between the behavior of living bodies and non-living matter find no place, and the test of all conclusions is experiment.

The factor common to alchemy and chemistry, is that of technique. The alchemists were the first and, before the latter part of the sixteenth century, almost the only laboratory workers. They had developed a small-scale technique of separating and combining the constituents of bodies, and to their equipment and technical methods, chemistry added almost nothing until it initiated the technique for collection and study of gases in the middle of the eighteenth century.

The transformation of alchemy to chemistry began with the passing of this technique into the hands of those who had objects other than the perfection of matter.

First of these, in point of time, were the pharmacists. From perhaps 100 A.D. to the thirteenth century, distillation was, almost exclusively, an alchemical practice. Though the pharmacists of Islam speak of distillation, they often mean a mere extraction of juices without vaporizing or condensing, and the instances where they conduct even a simple distillation are few. This was quite transformed when the distillation of spirits for medical purposes began, as described in Chapter IX. Many books on distillation appeared in the years after 1500, and they show us that stills—and moreover very complicated forms of stills—were being used by the pharmacist, and soon after by the housewife, for making all manner of cordials and distilled waters. It was quite obvious that the technique of alchemy was ministering to practical needs.

These distillation books brought alchemical technique into the home. Naturally the contact with the remarkable transformations described in these books interested the ordinary man in what could be done by natural science and natural magic, which were not very clearly distinguished. So, soon after the distillation books attained their full popularity, we find numerous books of recipes and secrets, giving instructions for all manner of "operative Distillations, Perfumes, Confitures, Dyings, Colours, Fusions and Meltings." Typical of these is *The Secrets of Master Alexis* (1555) which gives us recipes headed, e.g., "A very exquisite sope made of divers things"; "To make the hair yellow as Golde" (by a dye based on rhubarb root); "an oile or liquor to make the hair fall off" (soda, lime and orpiment); "To make a great deal of Incke quickly and with little cost." But with these household matters are very clear and practical recipes for chemical operations such as subliming calomel (used apparently as a face powder), for making vermilion, nitric acid, etc., and also workshop recipes for casting metals, gilding and so forth.

Yet there immediately follows this remarkable instruction: "*To take Salamanders for to occupy or to serve a man's turn.* When you see the Salamanders lie and sleepe in the Sun put on a pair of gloves, and so go to take them fayre and softly before they cast their venim (which is yellow). Then put it in some vessel of glass wherein there is man's

192

blood. Then it will serve your turne very well." For what purpose it was to be used we are not told—doubtless no good one.

These recipe books, which first appear about 1550, gradually degenerate. In the seventeenth century we find one opening with "How to have deinty sport with a catte"—far from dainty we should call it today. The main importance of these books was that they gave the general public the idea that natural science and especially the chemical arts were likely to be *useful*.

Technical treatises on chemical matters date back at least to the Assyrian chemical tablets (p. 18). In the Middle Ages and earlier these consisted as a rule of isolated recipes, rather like those of a cookery book, but from about the middle of the sixteenth century we meet with numerous treatises on the technical arts—metallurgy, glassmaking, fireworks, pharmacy, and the like. These generally have little or nothing of theory, but give excellent accounts of practice and constitute some of the earliest records of conscientious scientific observation. Biringuccio's *Pirotechnia* (1540) is one of the earliest among them, but the finest example is the great *De re metallica* of Agricola, published in 1556. It is impossible to give an account of this magnificent work which deals with every aspect of mining and metallurgy and includes a great deal that can be called industrial chemistry. In it we find the beginnings of chemical analysis in the assaying of ores, involving the use of small laboratory furnaces and delicate assay balances. There is the beginning of chemical manufacture in the preparation of sulphur, bitumen, saltpeter, nitric acid, vitriol, and the like.

Less considerable, but not unimportant, is the glassmaker's book *L'Arte Vetraria*, written by Antonio Neri and pub-

193

lished in 1612. Neri gives very clear directions concerning the way to purify his alkali and to select and pound his quartz pebbles so as to get a really colorless frit. He describes the removal of the green tint from glass by means of manganese

Fig. 27.—Assay balances, from Agricola's *De re metallica*.

and also the methods of making all the colored glasses, including ruby glass made from gold. He takes remarkable pains to recrystallize his materials. He tells us how to make nitric acid and *aqua regia* (nitro-hydrochloric acid) and uses the acid to dissolve metallic salts in order to incorporate

194

them with the frit. The beautiful medieval glass of our church windows witnesses the high technique that existed long before the time of Neri.

The typical contribution of the sixteenth and early seventeenth centuries to the formation of chemistry is, not new discoveries, but the recording, as a matter of general interest, of what had been the trade secrets of master craftsmen. We note the appearance of books about techniques which we should regard today as departments of industrial chemistry. The idea of "chemistry"—a science concerned with all the transformation of one substance into another—had not occurred to anyone, and the word *chemia* at that time meant simply "alchemy."

Herein was the chief contribution of that extraordinary man, Paracelsus. Philippus Aureolus Theophrastus Bombastus von Hohenheim, who appears to have assumed the name *Paracelsus* as an expression of his eminence, was a curious and violent character. He was born in 1493 and, early in the fifteenth century, studied first at the University of Basel, then under the mystically and magically minded Trithemius, Abbot of Spanheim. Later he went to the mining districts of the Tyrol where he made a study of minerals, mining, and the diseases of miners. In 1526 he became physician to the town of Basel and lecturer in medicine at that university. He violently attacked the then accepted theories of medicine and their holders and was involved throughout his life in perpetual quarrels. He wandered through Germany and Austria practicing medicine and writing his treatises. It is difficult to know whether to believe his enemies who accuse him of perpetual drunkenness and debauchery, nor can we be certain of the manner of his death in 1541.

195

At first sight the works of Paracelsus seem even more strange than those of the alchemists, but closer examination shows two important new factors.

First of all, we find a change of purpose. The Greek and most of the Western alchemists had been entirely concerned with the making of precious metals. Some of the Arabic writers, such as al-Rāzī, had concerned themselves with medicine, and the Lullian school of writers on the quintessences stressed their supposed medicinal virtue. But Paracelsus is the first to turn his attention wholly in the direction of healing. The object of all his discussions and recipes is the curing of disease, and only perfunctory mention is made of the philosopher's stone and the making of gold. His conception of nature is almost entirely a spiritual one, and perhaps his principal idea is the existence of quintessences in things, of an activity that can be separated or at least concentrated, so giving a particularly active medicine. The human body and each of its organs was supposed by him to be activated and guided by an "archæus" which was a spiritual being and which was influenced by the heavenly bodies, which were of the same nature. The quintessences, arcana, and other medicines that he tried to make were likewise spiritual, being full of the fifth element and so adapted to bring the heavenly influences to the archæus. Much of this doctrine is to be found in the Lullian treatises (cf. pp. 116-121).

Each of the seven metals, we have seen, corresponded to one of the seven planets and so the preparation of the quintessences of the metals for use as medicine was one of his chief aims. Quintessences are prepared by distillation, but metallic compounds are, with few exceptions, not volatile, with the result that the quintessences of Paracelsus, fortunately for the patient, contained normally none of the metal from which they were named. The method of making them was, in outline, to dissolve the metal in some chemical reagent such as nitric or hydrochloric acid, and to distil. The result was merely a rather dilute acid, the administration of which presumably did neither good nor harm. None the less, Paracelsus started the investigation of the use of metallic compounds in medicine and thereby gave a new impulse to the chemical worker whose energies had previously been confined to the well-worked but not overprofitable fields of alchemy.

In the second place Paracelsus had the first dim notions of the idea which we attach today to the word "chemistry." He himself retained the word "alchemy" but enormously extended its meaning. He sometimes speaks of it as the art of separating the "pure from the impure," and occasionally applies it to any work in which the potentialities of a material are brought into action. In some cases these definitions coincide. Thus, he says, God makes the medicine, but not in its prepared form; for, as it is found in nature, it is mixed with "dross," which must be taken away and the medicine set free. He had of course not the clear idea of this process that the modern chemist has. Today we know, for example, that there is a small proportion of a physiologically active chemical compound, quinine, in the bark of the cinchona tree and that this compound can be separated in a pure state and then purified no further. But if cinchona bark had been known to Paracelsus, he would have thought it contained a "quintessence of bark" possessing all the antifebrile potency of the drug, and which could be raised to any degree of potency by successive purifications, becoming more and more spiritual and volatile as the work proceeded.

The Alchemists

Since Paracelsus took this wide view of alchemy, he declared that without alchemy none could be a physician. But he stretches the term very much wider still. He gives the name of alchemist to the smelter of metals, the baker, the cook, and even to the man who lights and tends the fires. He gives to the word "alchemy" in fact very much the same scope as we give to the word "chemistry."

The actual chemical discoveries of Paracelsus were not considerable, nor did he contribute much that was of value to chemical theory. He and his followers introduced, in place of the Aristotelian four elements of earth, water, air, and fire, their "three hypostatical principles," mercury, sulphur, and salt. There was, of course, nothing new in taking mercury and sulphur as principles, for the notion that these were the principal constituents of metals was familiar in alchemy, being implied in the Greek texts, explicitly avowed by the Arabs and familiar in all Western alchemy. On the other hand, the introduction of salt as a principle seems to be new.

These three principles were not what we today know as mercury, sulphur, and salt. Thus the *Tyrocinium* of Beguinus (1611), which had 60 editions in 50 years, tells us that:

Mercury was that sharp permeating ethereal and very pure fluid to which all nutrition, sense, motion, power, colors and retardation of age were due. It was derived from air and water: it was *pabulum vitae* (the food of life) and the instrument nearest to form.[1] (Of this Boyle says, "this is not a definition but an encomium.")

[1] I.e., the chief agency by which changes were brought about.

198

From Alchemy to Chemistry

Sulphur was that sweet oleaginous and viscid balsam conserving the natural heat of the parts, instrument of all vegetation, increase, and transmutation and the fountain and origin of all colors. It was inflammable, yet had great power of conglutinating extreme contraries.

Salt was that dry saline body preserving mixtures from putrefaction, having wonderful powers of dissolving, coagulating, cleansing, evacuating, conferring solidity, consistency, taste, and the like. It resembled earth, not as being cold and dry, but as being firm and fixed.

These principles could not be separated from nature in fact, but they corresponded to what chemists thought they found. In distillations of organic matter the first volatile runnings were a "mercury," then came an oily "sulphur," a "salt" could be extracted from the dry residue. The notion of this mercury was still akin to that of the celestial virtue and of the philosopher's mercury. The three illustrations, Plates X to XII, show the way in which a fairly early Paracelsan, Leonhardt Thurneysser zum Thurn, symbolized the three principles and afford a useful corrective to those who would try to force on them too precise a chemical meaning.

The three principles of Paracelsus and his followers had almost all the vices of the four elements. They gave rise to a notion of matter, richer perhaps, but more confused and bound up with the occult notions of subtle matter. Yet the Paracelsan long remained a popular theory of matter, being later modified to the five principles, *phlegm, mercury, sulphur, salt, and earth.* It was held, in a form, by Becher, thence influenced Stahl and the phlogistonists; the phlogiston [2] of the eighteenth century is indeed the direct

[2] Cf. p. 211.

199

descendant of the "sulphur" of the Paracelsans which in turn
is the descendant of the "fire" of the Aristoteleans.

The great idea, then, of a single science, embracing all the
departments of sixteenth-century knowledge that we should
call today chemical, was due to Paracelsus, but it was long
in having its effect, because the theories of alchemy or
chemistry that were adopted by Paracelsus and his followers were ill-adapted to the explanation of the technical
practice of pharmacy, metallurgy, or any useful art, and
could not serve to unify them. The common factor in these
was the technique, and the thing needful for presenting
them as one science was the emphasis of the common elements of technique, as distinguished from theory.

The man who initiated this needful re-forming of the
subject was Andreas Libavius, whose chief work, *Alchemia*,
was published in 1597. A fuller version of its title is "Alchemy, collected by the labors of Andreas Libavius, Medical Doctor, Poet and Physicus, from the scattered works
of the best authors . . . and brought into an integral body."
There was as yet no possibility of classifying chemical
products according to their composition, which remained
largely unknown, and so Libavius arranged his work according to the type of chemical operation and final products
—not, as we do, according to the material in question. The
work is divided into two parts, (1) *Encheiria* (manipulation) and (2) *Chymia* (which consists of a classification of
chemical products according to their method of preparation
into magisteries,[3] extracts, distilled products, sublimates,
etc.). Libavius thus regarded chemistry as a practical art,
as is shown by his definition of "Alchemia" as *the art of*

[3] Drugs compounded from a number of simpler substances.

perfecting magisteries and extracting pure essences from mixed bodies by separation of their matters. His emphasis throughout is on the medical importance of the work.

He first proceeds to describe the instruments, glasses, furnaces; then the operations, such as calcination, incineration, sublimation, coagulation, fermentation, and the like. Then, in his second part, he turns to the varieties of products, such as potable metals, e.g., solutions of ferric acetate, mercuric nitrate ("but none will drink of this who is wise"), potable sulphur (a solution in turpentine); salts; amalgams; calces; crocuses; liquors, etc. The recipes are genuine, and in most cases it can be seen that some real chemical change is described or intended.

The work contains almost no chemical theory and in that sense cannot be regarded as an essay towards a complete textbook of chemistry, but it must have been a most valuable instrument to those who wished to acquire the practical technique. What little theory appears therein is still that of the alchemists. Thus: "A metal is a mineral body, constituted by the force of the seminary virtue of metals in the mineral ore of a vitriolic earth, from mercurial juice and sulphureous spirit, the vehicle of a digesting heat, and brought to the form of a fusible and malleable substance." Moreover Libavius believes in the philosopher's stone.

The pattern of the *Alchemia* was followed by the textbooks of the seventeenth century. It is not until the eighteenth century that we find the beginnings of the modern presentation in which the classification is by chemical composition. The preparation and properties of each substance are discussed in them in the light of principles laid down in the earlier part of the work.

Thus one of the favorite textbooks of the seventeenth

century was that of "Nicasius le Febure, Royal Professor in Chymistry to His Majesty of England and Apothecary in ordinary to His Honourable Household, Fellow of the Royal Society." This was first written in French and translated in 1670. "It contains whatsoever is necessary for the attaining to the Curious Knowledge of this Art Comprehending in General the whole Practice thereof." This is really a handbook for apothecaries and not a scientific study, but it starts with a theoretical introduction. The five principles, phlegm, mercury, sulphur, salt, and earth, are adopted. He distinguishes pure and impure substances: "by purity we will understand all what in the mixt or compound will serve our purposes . . ." His account of practical operations is admirable but smells of the apothecary's shop. His book includes all the elements of chemistry, theory, description, and practice, but is deficient in that the first of these is scanty and throws very little light on the rest.

Another similar and very popular work was Nicolas Lemery's *Course of Chymistry*, written and translated in 1677. Five pages suffice for the principles of chemistry; the remaining three hundred deal with practice. "Chymistry" is still the "art of separating mixts." The work is, however, better classified than that of Libavius. It is divided into mineral, vegetable, and animal chemistry, the germ of our separation into organic and inorganic chemistry. The compounds of each metal are grouped under that metal in a separate chapter, and the treatise does not look so very unlike a modern textbook of practical chemistry.

We have seen how the textbook of *practical* chemistry developed from the alchemical and technical traditions. The true *theory* of chemistry, on the other hand, was quite divorced from the alchemical. In the sixteenth century much thought was given to the mechanism and process which lay behind chemical changes, and we find in many authors who were not alchemists the developments of what were really alchemical ideas. One of the first of these is Bernardino Telesio who, in 1565, published a work on *The Nature of Things*. Like all his contemporaries, he failed to resist the temptation of constructing a complete system of the world from very inadequate materials. But he says very soundly that *the world is not to be inquired into by reasoning, but perceived by the senses directed to the things themselves.* He questions the need for the ideas of matter and form, elements and mixts. Heaven and earth are the only elements, and earth is transmuted by the sun's power into minerals, metals, juices, vapors. Transmutation, not combination or mixture, is his explanation of chemical change.

A somewhat similar system was held by William Gilbert, the writer on the magnet, who died in 1603. In his little-known and posthumous work, *A New Philosophy of our Sublunary World*, he says that the existence of elements is a fable. The action of the sun on the earth generates all. Nature does not make a compound by mixture like a pudding, but by growth like a plant; minerals *grow* from the juices of the earth, the great seminary. Nature makes bodies, not elements and mixtures. All is guided by natural attraction. Gilbert's is a fine work, but contains a great deal of theory to very little fact.

Anselm Boëtius de Boodt takes a rather similar view in his work on gems (1609). The earth *grows* into a stone or a gem. The proximate efficient cause of the change is a "lapidific spirit"; the remoter efficient cause is the celestial heat that puts the spirit into act; the remotest cause is God, *Deus Optimus Maximus*, parent of all things.

If these and the Paracelsan school of chemists had been the only contributors to chemistry, we might have said that alchemy absorbed the other aspects of the knowledge and practice of the changes in matter and so became modified into chemistry; but in fact there was another and very important school of chemical thought which was not alchemical at all. The ancient notion, that all bodies consist of atoms, though never altogether forgotten, made little or no appeal to the alchemists, for almost the whole of their theory is written around the Aristotelian ideas of matter. I do not say that a theory of transmutation could not have been based on atomic ideas; thus, for example, Plato's notion of earth, air, fire, water, and the celestial element being composed of atoms having the form of the five regular figures, allowed for the rearrangement of the triangles of air, fire, and water to form new atoms. The fact remains, however, that alchemy was actually based on the continuous theory of matter, and the revival of atomism started a totally new current of thought. Furthermore, alchemy could scarcely be separated from the idea of matter and form. (pp. 7-9), and determined attacks were made on this ancient doctrine.

One of the earliest atomists was Giordano Bruno who, in 1590, wrote his work on *The Principles, Elements, and Causes of Things*. His first principles are intellect and soul, above which is the absolute mind or truth. The material elements are earth and water, the immaterial ones, spirit and soul; in the material, are darkness, in the immaterial, light; from light and water proceeds fire. Air and spirit are aspects of the same thing. Light is a substantial spirit. Compounds

204

are formed by mixture of bodies but are composed of atoms. The scheme is obviously not very different from the alchemical system. There is a great deal in the work about magic and Moses, and one feels that, if Bruno had not qualified as a hero of rationalism by being burnt for his theological errors, posterity might have called him superstitious.

From about 1620 the *atomists* become important. Francis Bacon did not write much concerning the elements, but was an atomist with a clear idea of the importance of the fine structure of matter. He regarded the element of fire as a figment and considered the essence of heat to be motion. He still retained the Hermetic idea that the beautiful and elaborate works of the crust of the earth arise from the influences and perpetual animation of the celestial bodies.

Galileo Galilei (1564-1643) continually speaks of the motions of the particles of bodies and regards heat as a swarm of tiny corpuscles penetrating into bodies. But he was not much concerned with the nature of terrestrial matter, though his demonstration that the heavenly bodies appeared to be of terrestrial material was evidence against the existence of that heavenly matter upon the existence of which much of alchemical theory depended; it struck furthermore at the root idea of the alchemical world, the life of the heavenly bodies.

In passing we may note Sebastian Basso who, in 1621, wrote twelve books of philosophy against Aristotle. He attacks the doctrine of matter and form. If the substantial form of man gives him his properties, does the form of man breed lice?—he asks. But actually he makes little advance towards the theory of chemistry. He retains three of the four elements, though in place of the element of fire he sets

205

a "spirit" consisting of tiny needle-like particles penetrating everything. The idea of this spirit was akin to our idea of energy; its function was to bring about the combination of elements into compounds.

These tentative essays towards atomism were eclipsed by René Descartes. He was in no sense a chemist, but since he put forward the first systematic atomic philosophy of modern times, he thereby influenced every subsequent author. In his *Principles of Natural Philosophy* (1644) he sets out his system. Matter is atomic, atoms are simply extension; extension and motion constitute and explain all phenomena. The question of a prime matter therefore does not arise. Like all early atomists, he thinks of his atoms as differentiated into the customary small number of elements. "The first kind is that which has so much force of agitation that, by entering into other bodies, it is divided into minutiae of indefinite smallness and accommodates its shape so as to fill the narrowest corners. The second, is that which being divided into spherical particles, very minute . . . yet are of a certain and determinate quantity and divisible into others yet smaller . . . a third, consisting of particles either more coarse or less adapted to motion. The sun and stars are composed of the first, the heavens of the second, the earth and planets of the third." Thus Descartes really retains the notion of "spirit" in his first and second matter, but then takes the important step of disassociating them from mind which he regarded as wholly immaterial.

This hypothesis, though based on the flimsiest evidence, was at least a physical and mechanical explanation. Descartes applied his atomic theory to explaining chemical phenomena, but not with great success. Here is a fair sample:

From Alchemy to Chemistry
WHY SPIRIT OF WINE VERY READILY BURNS

Indeed spirit of wine very easily nourishes flame because it consists wholly of very slender particles and on these there are certain little branches, so short and flexible indeed, that they do not adhere to each other (for then the spirit would be turned into oil) but such as may leave very small spaces about them which cannot be occupied by globules of the second element,[4] but can only be occupied by the matter of the first element.[4]

Descartes is followed by several other atomists who theorized, but made little attempt to relate their theories to chemical observations. J. C. Magnenus (1648) in his *Democritus reviviscens* ("Democritus coming alive again") relates the doctrine of elements to that of atoms. Elements consist of atoms of the same kind and similarly figured. He still thinks there are only three elements, earth, water and fire, and ordinary bodies are mixtures of these. Matter and form he regards as a way of thinking, not as really existent in nature.

Much better known is Peter Gassendi who, in 1649, published his *System of the Epicurean Philosophy*. Matter is atomic, the atoms have shape and size, and all consist of the same material; they are indivisible on account of their solidity. He speaks not only in terms of atoms, but also of molecules (*moleculae*), a notion which, however, is to be found in the works of the ancient Greek atomists. "There are molecules or, if you like, very tiny little concretions which, being made by certain types of more perfect and indissoluble coalitions, long endure as the seeds of things, which are not atoms, but these that can be resolved into atoms."

[4] René Descartes. *Principia naturae.* Amsterdam, 1644, Part IV, Ch. ciii, p. 246.

The Alchemists

Next follows Robert Boyle, who was the first to make an attempt to form a theory of matter directly useful to science. But we may mention in passing one whom Boyle had read, that godly and learned man, Mr. William Pemble, whose work, *On the Origin of Forms* (1639), was dedicated to Accepted Frewen (President of Magdalen). He doubts whether substantial forms exist save in the heads of philosophers, and he argues that a form is not a substance. He regards bodies as accidents inherent in prime matter, i.e., independently alterable properties inhering in the same matter. He still retains the celestial virtue and the elements of air, water, and earth; but his attack on forms was noted by Boyle.

The growing impatience of the time with Aristotelean and scholastic modes of thought on such matters finds witness in Samuel Butler's *Hudibras*, "written in the time of the late warres." His hero was an adept therein:

> What ever Sceptic could enquire for,
> For ev'ry *why*, he had a *wherefore*.
> . . . His Notions fitted things so well,
> That which was which he cou'd not tell:
> But oftentimes mistook the one
> For th'other, as great Clerks have done.
> He cou'd reduce all Things to Acts,
> And knew their Natures by Abstracts;
> Where Entity and Quiddity,
> The Ghosts of defunct Bodies fly;
> Where Truth in Person does appear
> Like Words congeal'd in Northern air.
> He knew *what's what*, and that 's as high
> As Metaphysick wit can fly.
>
> (*Hudibras*, Part I, Canto I, l.131)

From Alchemy to Chemistry

The world was rapidly ceasing to have interest in *what's what*, in essential being, and was ready to turn its philosophic minds to what the older philosophies had thought so unimportant—particular changes in individual things.

From 1661 onwards, Robert Boyle showed the unsatisfactory character of the doctrine of forms as applied to particular phenomena, and the entire inadequacy of all theories of elements and mixtures as then held. This work is chiefly contained in the *Sceptical Chymist* (1661) and the *Origin of Forms and Qualities* (1667). Boyle's own hypothesis concerning matter was an atomic one, not much different from that of Gassendi. The material world consists of atoms and clusters of atoms in motion, and every sort of phenomenon is brought about by one particle hitting another. The idea of forces acting at a distance did not become significant until it was developed by Newton.

Boyle attacks the older doctrine by appeal to *experiment* and to *metaphysical argument*. Thus experiment shows that all bodies are not resolved into the same few elements, and that the supposed earth, water, air, and fire, obtained by breaking up bodies by heat, are neither elementary nor identical when made from different bodies. He concludes that there is no *fixed and determinate number of elements* into which fire resolves bodies. He doubts, indeed, if there is any proof of the existence of elements, though he is far from denying it. He does not, however, propose any practical method of discovering whether a body is an element, nor does he make any list of elements. Consequently, his ideas on elements remained substantially barren until Lavoisier established a clear and definite doctrine of elements derivable from experiment. In the *Origin of Forms and Qualities*, Boyle attacks the whole idea of a substantial form abiding in matter and causing its properties. He concludes that such forms are unknowable and so useless in science. He does not entirely reject the idea of matter and form but says:

The forme of a natural body being, according to us, but an essential modification, and as it were the stamp of its Maker, or such a convention of the bigness, shape, motion (or rest) situation, and contexture (together with the thus resulting qualities) of the small parts that compose the body, as is necessary to constitute and denominate such a particular Body, and all these Accidents being producible in Matter by local motion . . . the first universal though not immediate cause of formes is none other than God . . . and among second causes, the Grand efficient of Formes is *local motion*.[5]

Thus Boyle would say that iron is hard, not because the substantial form of the iron is the cause of the hardness in it, but because the shape and motion of its parts render them deformable only with difficulty. Thus, since Boyle, we have been able to hold a metaphysical theory of matter, if we so desire, but independently of any such theory to explain physical and chemical phenomena by means of local motions, velocities, forces, etc. Even Boyle's version of the atomic theory was not of much value in bringing about fruitful explanations of matter, though it was used by Mayow in his admirable *Five Treatises*. Boyle's sound views on elements had but little effect, for the chemists of the eighteenth century still thought in terms of spirits, earths, and other entities recalling the old Aristotelean elements. There was not much talk of elements and atoms, indeed, until Dalton related the idea of atoms to the quantitative relationships of Lavoisier's elements.

[5] Robert Boyle, *Origin of Forms and Qualities*. Oxford, 1667, p. 101.

From Alchemy to Chemistry

We have traced the development of works on chemical practice and on chemical theory, but we have not yet found them combined in a single textbook of chemistry.

The first balanced combination of chemical theory and practice seems to be found in Hermann Boerhaave's *Elementa Chemiae* (1732). For him Chemistry is an *art which teaches the manner of performing certain physical operations and also seeks to investigate their causes.* The familiar form of the chemical textbook with its preparations and lists of properties begins to emerge, though it is still embedded in a great deal of discursive writing. Yet, his work still contains the Aristotelean idea of the elements and the notion of imponderable principles, so that his chemical theory did very little to assist the practical operations.

Indeed, the combination of true theory and sound practice, which was needed before chemistry could advance with speed and confidence, was not realized until the last of the alchemical ideas, that of phlogiston, was exploded. The theory of phlogiston was the standard explanation of combustion throughout the eighteenth century. In brief, it was supposed that a body was combustible because it contained the material principle of combustibility—phlogiston. The combustion of a body was the streaming-forth of phlogiston. Bodies which promoted combustion were those that lacked phlogiston and could, therefore, readily receive it from the combustible. This idea of a principle of inflammability common to all inflammable bodies is, of course, very ancient. It is Aristotle's element of fire, the alchemists' sulphur. When it was first proposed by J. J. Becher it was supposed to be a "fatty earth," but many later phlogistonists treated it as an imponderable fluid. It was Lavoisier, whose *Révolution chimique* finally cleared out these last vestiges of the ancient modes of thought, who gave to chemistry a sound foundation which has never needed to be rebuilt.

Thus the only part of alchemy which formed a permanent part of chemistry was its laboratory technique. Its theoretical part formed a temporary means of fitting chemical change into the natural philosopher's world. But as scientific method reformed chemical theory, the specific ideas of alchemy were not so much disproved as found useless and discarded. The notions of the correspondence of chemical operations with the heavenly bodies, of the analogy between chemical changes and those of living beings, were not found useful by experimental chemists and gradually died out. The notion of "spirit" lost its psychic quality, and although the various "subtle matters" of the seventeenth and eighteenth centuries (e.g., ether, electric and magnetic effluvia, caloric, the animal spirits) were really descendants of the ancient *pneuma*, they were not regarded as living or akin to mind. Today we have lost the last survivor of these—the ether of space; our world, as seen by science, is now wholly impersonal and nothing akin to mind.

XIV

The Hermetic Philosophy

THE last chapter dealt with the manner in which chemistry took shape as something different from alchemy, different in its purpose, its method, and its technique. The purpose of chemistry was to investigate the different kinds of matter and their changes, to accumulate in the Baconian manner a natural history of bodies, from it to elaborate a natural philosophy with the help of "experiments of mechanic art," and at the same time to provide clear directions for the best methods of making the various substances required for the practical arts. The essential purposes of alchemy, namely, the perfection of matter and the understanding of it in terms of a spiritual world, were totally absent from chemistry. The method of chemistry was experimental, the recording of the changes caused by laboratory treatment and the drawing of hypotheses to account for them. The essential method of alchemy, its interpretation of the ancients and consideration of nature through human symbols, had no part therein. The technique of chemistry was wide and varied and notably quantitative; it went far beyond the digestions and distillations of the alchemists.

No one could fail to see that chemistry was a means of

obtaining real knowledge about things; it did not promise the dazzling prospect of transmutation and the understanding of the ultimate nature of things, but what it did promise, it performed. Thus the man of the seventeenth century who had an interest in the changes that take place in the different kinds of matter was drawn to chemistry rather than alchemy. Such a man as Robert Boyle, though he did not regard transmutation as impossible, was wholly a chemist and in later life scarcely considered the claims of alchemy to be worthy of investigation.

But the growth of chemistry did not at once bring about the end of alchemy, for in the seventeenth and early eighteenth centuries a vast amount of alchemical literature was published and eagerly read. But the alchemists' ranks no longer contained those who had a habit of mind, scientific in the modern sense of the word. Such men became chemists. The alchemist of the late seventeenth century was one who was not so much interested in the details of laboratory processes as in the essence or being of matter, its relation to man and its purpose in the universal scheme. Thus the character of the alchemical books steadily changes from descriptions of the practical arts to the setting out of grand but cloudy schemes of the universe as seen under a spiritual guise.

Why were such works as these so popular at that time? The seventeenth-century inquirer into nature had been educated in the idea that the aim of "natural philosophy" was to give a complete account of the natural world. There was no doubt in his mind that the world had been made by God and was kept in existence by Him for a specific purpose and in a manner literally or figuratively indicated in the Holy Scriptures. God, and man in part, were superior to, but not separated from nature, and like the rest of nature, must enter

CORPVS SAL

NE PRAETER ALIAM

PLATE XII. SALT
(From Thurneysser's *Quinta Essentia*, 1570)

PLATE XIII. DISTILLATION ON THE LARGE SCALE, c. 1500

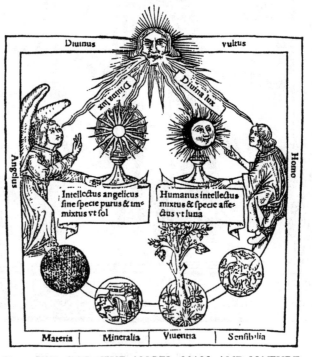

PLATE XIV. GOD, THE ANGEL, MAN, AND NATURE
(From the *De Intellectu* of Bovillus, 1510)

into natural philosophy. The seventeenth century, like those before it, was still imbued with the idea that nature was a type or picture of the Divine, that in her workings were to be seen the workings of God, and an account of the world that left out the activity of the Author and Preserver of nature was simply not a natural philosophy. But the new science, which was then becoming of interest, did leave out all that related to God or to the individual human observer, not designedly, but simply because it considered nothing but such relationships between classes of individual things as could be elicited from the observations that anybody could make and communicate fully to anyone else.

The new science asked and answered questions as to how the stars moved, how bodies fell, what new bodies were formed by chemical operations; but it never asked and could not ask the reasons for the order that was discovered, nor could it connect what it described in nature with what man experienced in his personal relation with nature and with God.

The medieval philosopher could visualize the whole cosmos with the vast empyrean heaven enclosing the concentric spheres of the planets which, in their turn, governed all the changes of the world. He saw these changes as operated by God's will, doing God's purpose. He saw the world as begun by God and by Him to be ended. The new science left out all this, and consequently it seemed to the philosophical and religious thinkers to be lacking in interest or at least to be insufficient. It revealed a number of instances of law and order, no doubt. But was a man to renounce this wonderful vision of a world impelled by God for God's purpose in order to trifle with the measuring of pendulums and the weighing of air?

The Alchemists

The great majority, whether of scientists or philosophers, found this renunciation repugnant and impossible, and were consequently involved in the difficulties of combining the religious account of the world with the scientific one. Galileo, Descartes, Boyle, Newton, Leibniz, all had to seek their harmony between what has been called the *vertical* world view, in which all descends from God, the Father of Lights, to the world He created, orders and preserves, and the *horizontal* world of observed relationships between material objects. It would be a long task to chronicle the solutions that were given. To put the matter briefly, however, any theory that regarded matter as in no way living, completely sundered from mind, and mind as wholly immaterial, completely sundered from matter, found difficulty in explaining the interaction of matter and mind in man, and likewise the governance of the world by God. Yet, despite the difficulties it raised, the general trend of the scientists was towards this severance of mind and matter, and this is still the typical attitude of present-day science.

Now, the view of the world that was held by the older natural philosophers, and which was essential to alchemy, proposed as the principal agent of nature a *middle* substance, mediating between mind and matter, and so being capable of resolving many of these difficulties. The medieval scheme of the world could no longer satisfy the men of the seventeenth century, because the bottom had been knocked out of it by such innovators as Galileo, Boyle, and Newton, who showed that much that was essential to it was not true. But nothing had been said to upset the belief in our *pneuma*, the middle substance, and, under the guise of the animal spirits, the magnetic effluvia, the ether of space, etc., something very like it still played an important part in scientific theory.

Accordingly, writings which tried to resolve in terms of *pneuma* the difficulties that a materialistic science was raising were sure of a hearing. Behind alchemy, as we have seen, there was always a theory of the operations of nature, and this theory, in somewhat revised forms, was presented anew to the seventeenth-century public under the name of the Hermetic philosophy.

The name of the Greek god Hermes, as we saw, was attached to a number of writings concerned with religion, astrology, magic, and alchemy in the first few centuries after the beginning of the Christian era. But the Hermetic philosophy of the seventeenth century had many other sources. The Neoplatonists, with their theory of the emanation of light from God and its descent into and animation of matter, were of equal importance and, of course, the results of the interpretation of the Scriptures were fundamental.

The general outline of this account of the world is found in the works attributed to Ramón Lull as early as the fourteenth century, and doubtless much of it is of Arabic origin. In the fifteenth century, first in Italy and later in Germany, there was intense speculative activity concerning the nature of things and a secret breaking away from the somewhat stereotyped official medieval cosmology, especially in the direction of the study of "magic." This magic was of two chief kinds, "black magic," the invocation of demons in order to compel or persuade them to perform preternatural works, and "natural magic," the discovery of supposed hidden relationships which could be used for such purposes. Of this latter kind were operations with sigils, talismans, gems, herbs, etc., and the line between natural magic and natural science was in the sixteenth century very indistinctly drawn.

217

The Alchemists

The practice of black magic was, of course, forbidden by the Church and severely repressed. The position of natural magic was uncertain, for its exponents avowed it as an altogether admirable and pious practice, while the Church viewed it with suspicion. But in countries where the power of the Church was weak or non-existent at that time, as in parts of Germany and England, natural magic was openly discussed and practiced. This natural magic was very dear to the heart of the ordinary man of the seventeenth century. It is a paradox that this century produced both the men who were the fathers of the rationalistic attitude which finally made such magic incredible and, at the same time, a populace which had an inexhaustible appetite for such marvels. Never was there an age when the literate were so avid of the marvels of astrology and magic.

In the seventeenth century, not only was there a dissatisfaction with the new science, which was seen not to be a philosophy of nature as a whole, and which consequently awakened a desire for such a philosophy, but there was also a wish for an explanation of the world in which the surviving beliefs in the "Hermetic" sciences—alchemy, astrology, and natural magic—should find a rational justification. The result was a defense of these beliefs by explaining them and making them appear rational in the light of a spiritual natural philosophy, the Hermetic.

This philosophy appears in multitudinous versions between the time of Heinrich Cornelius Agrippa von Nettesheim (1486-1535) and the end of the eighteenth century. The most accessible English source is the writings of Thomas Vaughan ("Eugenius Philalethes"), brother of Henry Vaughan, perhaps the greatest of the English religious poets. Thomas Vaughan was born in 1622, studied at Jesus College, Oxford, and about 1640 became the parson of St. Bridget's, Breconshire. He was obviously deeply impressed by the beauty of the Welsh countryside, and between 1650 and 1655 he published his remarkable works on the Hermetic philosophy. He certainly practiced as an alchemist, but, it seems, abandoned the laboratory practice in favor of his philosophizing. His work shows everywhere an intense love of nature and a desire to fathom the secrets of her life.

The Hermetic philosophy was mysterious, both necessarily and deliberately. Necessarily, because it took account of God and an invisible world which could not be observed by the senses and, therefore, could not be described in a manner as to be visualized. Deliberately, because the knowledge, which the Hermetic philosophers believed themselves to be able to attain, would, they supposed, confer a degree of power which, in the hands of the wicked, would be disastrous to the world.

The essential idea of the Hermetic philosophy is the causal chain of descent from God to matter. Following the Scriptures, the origin of all things is believed to be God who first considered and so formed the eternal idea of all things. The goodness and beauty of this idea moved him to make a material copy. God, the Father, is the supernatural foundation or basis of his creatures; God, the Son, is the pattern in whose image they were made; and God, the Holy Ghost, is the spirit that framed the creation in due proportion to the pattern. God, the Father, is thus compared to the Sun; God, the Son, to Light; and God, the Holy Ghost, to a fiery love, a Divine heat. All of this is, of course, compatible with Christian theology.

The Alchemists

The process of creation of the world, as given in Genesis, is evidently the combination of two different accounts. We find in one the image of the Holy Spirit resting over the waters, but this "water" was evidently a created thing. God was supposed by the Hermetists to have created the "first matter," something that was not yet any particular thing, but that was potentially everything. This "first matter" was a horrible empty darkness which condensed into that primitive water of which the Scriptures speak. Light emanated from God (not mere physical light but the Word that is "Light," of which we hear in the first chapter of St. John's Gospel); it pierced the matter and formed in it a model or pattern which contained potentially all that was to be in the world. The Holy Spirit, working upon this chaos, this "huddle or limbus of all things" (and here is the typically alchemical side of this philosophy), separated the subtle from the gross by a sort of cosmic distillation or sublimation— "terrible mysterious Radiation of God upon the Chaos and dark Evaporations of the Chaos towards God."

In this way, by the divine heat and light, there was first separated from the mass a thin spiritual celestial substance containing three parts of light to one of matter. This became the most spiritual part of the world, forming the bodies of Angels, the empyrean sphere, and later the body of the sun and heavenly bodies. Vaughan calls this substance the *Anima*, and it is the same as the "argent vive" of the Lullian texts (p. 118). Next was separated from the chaos a less subtle substance, two parts of light to three of matter, the *Binarius*, which formed the interstellar heavens, and the still less subtle *Ternarius*, one part of light to three of matter, which is the *pneuma*—air, breath, spirit—of which we have heard so much. In this system, the *Ternarius* is the link between the celestial and the terrestrial world. It is the subject on which the alchemist and magician work, and the means of transmitting the influences which are the subject of astrology.

For Vaughan, at least, air is not an element, not a substance material in the sense that water and earth were material. It is a miraculous hermaphrodite, nature's commonplace. In it are innumerable magical forms of men, beasts, herbs, and trees; it is the receptacle of spirits after dissolution. It is the fuel of the vital sensual fire. It is the magician's back door, his fire that passes through all hands. In this air, or *Ternarius*, resides the secret of all the occult arts.

The residue of the original mass, after the extraction of these subtle bodies, consisted of the elements water and earth. These contain in them but little of the light, though every kind of body has a seed of it. Water—by which, of course, we do not understand the chemist's H_2O, but a principle of liquidity, fertility, and the like—is a more subtle creature than earth. Water can rise into air as vapor, and fall in dew. It is a link between air and earth and carries the celestial influences down to earth. Finally, there is earth which is the receptive part of the world, acting as the womb or matrix within which all generation takes place, receiving all influences.

In this system, as in many other systems of the time, there is no element of fire. Fire is not so much an element as an activity; it is that which descends from God, a moist, silent fire that operates all things in nature, the Cupid that weds the Psyche of water.

The elements were each threefold. There was, e.g., a spiritual earth, a celestial earth, and an elementary earth.

221

This constitution gave the terrestrial elements their power to correspond to the changes in the celestial world. Thus the spiritual earth corresponded to the changes in the active spiritual bodies, e.g., the sun; the celestial earth corresponded to the passive celestial bodies, e.g., the moon. There was in the terrestrial elements a little sun and a little moon which could marry and generate; these were the fire and the moisture, and they were influenced by the sun and the moon respectively.

Consequently, the generation of new things depends on the marriage of the fire and the moisture in the womb of earth. But *what* new thing was to be generated depended on a portion of the original light seeded in the substance in question, which Vaughan calls the "invisible central artist." This is what determines, e.g., that a lion shall beget a lion, and not a dog, or that gold should be begotten of gold, and of no other metal.

The actual theory of the process of alchemy remains somewhat obscure and, indeed, takes a minor place. But the stone is a condensed and tangible form of the light, obtained by a separation of the subtle from the gross. It is based upon the subtilization and subsequent fixation of the *Ternarius* and is the substance that contains the maximum of light. In this way it can perfect any body, for the light is the idea of the perfect world that God considered. Thus we can understand how the stone came to be symbolized by the figure of Christ and why it was regarded as His analogue in the inferior world. Christ was the perfect man and God, so constituted for the purpose that man should be redeemed and perfected. In like manner the stone was the perfect matter, made up of God's light and a spiritual body, by which bodies were to be redeemed and perfected.

It cannot be denied that this was at least a poetically inspiring scheme of the world. It bore no relation to physical science as we know it, but it linked God, man, and matter in one scheme; it showed all nature as God's handiwork, modeled by the seed of God's light within everything, energized by the continuous flow of influence from heaven to earth. It opened to man the possibility of *knowing* nature by the cultivation of his powers, instead of merely chronicling her external changes; it promised him the understanding, not so much of the reasons for phenomena, as of the life principle that lay behind them. It was a conception deeply inspiring to the religious and artistic side of man. Seventeenth-century poetry is full of its influence, and Vaughan himself is a poet of no mean order. See how he speaks of his *Hyanthe*, the moist passive nature that is worked upon by the fire from above:

HYANTHE

It was scarce *Day*, when all alone
I saw *Hyanthe* and her *Throne*.
In fresh, *green Damascs* she was drest,
And o're a *Saphir Globe* did rest.
This slipperie *Spheare* when I did see,
Fortune, I thought it had been *Thee*.
But when I saw shee did present
A *Majestie* more *Permanent*,
I thought my Cares not lost, if I
Should finish my *Discoverie*.

Sleepie shee look'd to my first sight,
As if shee had *Watch'd* all *the Night*,

223

The Alchemists

And underneath, her *hand* was spread,
The *White Supporter* of her *head*.
But at my second, studied *View*,
I could perceive a silent Dew
Steale down her *Cheeks;* lest it should *Stayne*
Those *Cheeks* where only *Smiles* should reigne.
The *Tears* stream'd down for *haste*, and all
In *Chaines* of *liquid Pearle* did fall.
Faire Sorrows; and more *dear* than *Joyes,*
Which are but emptie *Ayres* and *Noyse,*
Your *Drops* present a *richer Prize,*
For *they* are Something *like* her *Eyes*

Pretty *white* Foole! why hast thou been
Sulli'd with *Teares* and not with *Sin?*
'Tis true! thy *Teares,* like *Polish'd Skies,*
Are the *Bright Rosials* of thy *Eyes,*
But such strange *Fates* do them attend,
As if thy *Woes* would never end.
From *Drops* to sighes they turn, and then
Those *sighes* return to *Drops* agen:
But whiles the *Silver Torrent* seeks
Those *Flowr's* that watch it in thy *Cheeks,*
The *White* and *Red Hyanthe* weares,
Turn to *Rose-water* all her *Teares.*

Have you beheld a *Flame,* that springs
From *Incense,* when *sweet, curled, Rings*
Of *smoke* attend her *last, weak Fires,*
And *Shee* all in *Perfumes* expires?
So dy'd *Hyanthe.* Here (said shee)
Let not this Vial part from Thee.
It holds my *Heart;* though now 'tis *spill'd,*
And into *Waters* all distill'd.

224

The Hermetic Philosophy

'Tis *constant* still: Trust not false Smiles,
Who *smiles*, and *weeps* not, shee *beguiles*.
Nay trust not *Teares*: *false* are the *few*,
Those *Teares* are *Many*, that are True.
Trust *Mee*, and take the *better Choyce*,
Who hath my *Teares*, can want no *Joyes*.

       (*Magia Adamica*, 1650, pp. 93-95)

In a different vein, note the beauty of this poetical prayer,
phrased in the language of the Hermetic philosophy:

Lord God! this was a *stone*
as *hard* as any *One*
Thy Laws in Nature fram'd:
'Tis now a *springing Well*,
and many *Drops* can tell,
Since it by Art was tam'd.

My God! my *Heart* is so,
'tis all of *Flint*, and no
Extract of *Teares* will yeeld:
Dissolve it with thy *Fire*,
that something may aspire,
And *grow* up in my Field.

Bare Teares Ile not intreat,
but let thy *Spirits seat*
Upon those *Waters* bee,
Then I new form'd with *Light*
shall move without all *Night*,
Or *Excentricity*.

       (*Anthroposophia Theomagica*, p. 28)

The Hermetic philosophy, then, supplied a need that science could not satisfy, yet the number of those who felt that
need was small. For even in the seventeenth century, this

225

philosophy had but a limited success. The current of the world's thought had been fixed in a wholly different direction. Very little was heard of it in England after 1720, and even in Germany, where it had the greatest measure of success, it scarcely saw the nineteenth century.

As a matter of history, the Hermetic philosophy became more and more difficult to credit as transmutation and magic failed to manifest themselves to the new criterion of scientific inspection. It fell into final disrepute when the imponderable spirits—the effluvia, the electric fluid, the matter of light, phlogiston, caloric, the animal spirits, and the like—became discarded from science and man ceased to be able to think in terms of breaths and influences.

But even in the nineteenth century there was an aspect of alchemy that could rouse serious interest. It could no longer be usefully pursued by those who sought practical results or a philosophy verifiable in nature. Alchemy had, however, a significance different from either of these. From its earliest period the alchemical process had always been regarded by some of its exponents as obscurely setting forth a *mystical* process performed in the mind and intended to bring about the regeneration of man. Man, on this theory, was the source of the philosophic mercury, a view easily justifiable in the days when the agent of his will was taken to be the *psychikon pneuma* or "animal spirit," which view persisted in certain medical circles till the end of the eighteenth century.

Man was the alchemical vessel in which this spirit was to be elaborated. Man was likewise the base metal which was to die and be regenerated as gold. Such a belief is, I think, to be found in some early Greek texts, such as the *Dialogue of Cleopatra and the Philosophers* (pp. 57-59). It is clearly

226

present in the seventh-century work of Stephanus, and passages that can be read in this way abound in the alchemical texts of all periods.

Thus, the latest alchemical writers, who found the chemical aspects of alchemy both untenable and repugnant, attempted to show that alchemy was essentially a mystical process, and that those, past and present, who treated it as a process designed to make real gold were laboring under a vulgar error from which the true initiate had been delivered.

This view is, I think, untenable. The mystical interpretation of alchemy is a possible one and had in all periods its adherents, but the chemist who studies the alchemical texts cannot fail to see in them the fruit of laboratory work. If the materials, vessels, and methods of alchemy were mere symbols, standing in the same relation to chemistry as the symbols of freemasonry to architecture, then we should not have found the alchemists to be the inventors of chemical technique and the designers of apparatus which would be usable by the chemist of today. Moreover, if, as C. G. Jung seems to suggest, the alchemical phenomena were mere visions or projections of the unconscious upon the matter contained in the alchemical vessels, there is no reason why alchemical apparatus should be well-adapted for practical work with chemicals as we know them today.

We must, in fact, allow a primary practical tradition, but there is no reason to deny the existence of a school of mystical alchemists whose purpose was self-regeneration. It is, indeed, quite evident that the alchemical terminology was used in purely mystical writings as early as the sixteenth century. Jacob Boehme's [1] works, for example, are certainly mystical, and they use the words nitre, sulphur, mercury,

[1] German mystic (1575-1624).

salt, etc., to denote spiritual entities existent in man as well as in the great world; and anyone who regards Thurneysser's symbols for the three last named (Plates X to XII) can easily understand the possibility of so doing. No one, however, could mistake Boehme's work for alchemy; he is obviously wholly sundered from the laboratory. The Rosicrucian works treat alchemy in much the same manner, and in the eighteenth century we find many alchemical books which seem to be much more mystical than practical.

This tendency culminated in 1850, when Mrs. Atwood, formerly Miss South, in company with her father, Mr. Thomas South, wrote a remarkable work, which every student of alchemy should read, if he can obtain it, namely, *A Suggestive Enquiry into the Hermetic Mystery*. This book is deeply impressive and evidently written in a fire of enthusiasm. The style is strange and archaic, modelled upon that of Thomas Taylor, the Platonist. The book compares alchemy with the mysteries of the ancients, and maintains it to be a mental process. The first matter is the middle element of the Hermetic philosopher; the *Ternarius*, the region of man's fantastic and imaginative existence, the "astral" region of the theosophists; the vessel is likewise man, and the work is the purification and exaltation of this lower part of the mind so as to accomplish the mystical work and join man to God. The process she believed to be "manual," as the alchemical works so often say. But whereas the natural interpretation of this word leads us to think of alchemy as the laboratory manipulation of matter and vessels with the hands, as in chemistry, Mrs. Atwood equated the word "manual" to the use of the hands in the induction of the hypnotic trance, which was then newly discovered and of great interest.

The Hermetic Philosophy

She supposed that it is possible to influence and manipulate the lower part of the mind of man by means of the hand of the adept. This process was to enable the first matter to be elicited from man and to be used in the alchemical process which, as I read her, was likewise to be performed in man as vessel. I am myself not impressed by the evidence for the "manual" process. Furthermore, we have no reason to believe that Thomas South, Mrs. Atwood, or anybody else in modern times accomplished anything more than the mere induction of hypnosis by such means, and I am inclined to regard the chief contribution of the *Suggestive Enquiry*, deeply impressive as it is, to be the establishment of the existence of mystical alchemy, and to characterize its thesis, that alchemy is *essentially* a mystical process induced by a "manual" one, as an interpretation of alchemy which the alchemists of earlier times would not have recognized as true.

To treat alchemy as no more than plain material chemistry is undoubtedly an error; to treat it as no more than an interior mental process is no less.

If alchemy were identical with mysticism, it is hard to understand why the mystically-minded, who desired to transmute themselves, should have resorted to this jargon of sulphurs and mercuries, alembics and crucibles, at a time when the world's greatest mystics, Ruysbroeck, Eckhart, the author of the *Cloud of Unknowing*, and later St. Teresa and St. John of the Cross, were writing in a language the meaning of which was not intentionally hidden but only obscure through the nature of the subject. Obviously, the alchemical writings, if they were only a cover for some mental process, are unlikely to have been a cover for Christian mysticism, the pursuit of union with God.

It would appear that all the aspects of alchemy can be explained, if it is taken to be a practical natural philosophy, as it were, a chemistry of that entity "spirit" or "breath" which was believed by the alchemists and Hermetic philosophers to enter into both man and metals. This *pneuma* or spirit, the middle substance between celestial and terrestrial, is the essential material of alchemy, of whatever period. The earliest alchemists identified distilled water and "sublimated vapors" with this *pneuma;* later other volatile liquids, such as nitric acid and alcohol, seemed to be the *pneuma;* again it was identified with the magician's supposed instrument that Vaughan describes; finally it was thought of as in some sense a part of man's mind.

It is clear then that, while men believed in the *pneuma* as agent in nature and in man's mental operation, and called the manipulation of that *pneuma* by the name alchemy, this word could be applied to several different processes: first, to operations with distilled and sublimed substances in the laboratory, to something that we could call a material chemistry of volatile bodies; secondly, to a mental process in which a magician attempted to draw on the *pneuma* outside him and to project that *pneuma* into a vessel and fix it so that it could be handled as a material substance; thirdly, to manipulations of the *pneuma* or spirit of man, which could in fact amount to a truly mystical process. Let alchemy be called "a chemistry of spirit," and it will be possible to understand its many aspects and the conflicting views of those who have not grasped its essential features. .

XV

The Relation of Alchemy to Science

THE man of science who looks for the first time into an alchemical text expects to find something like a textbook of chemistry, though much less developed and accurate. But, in fact, he finds something that scarcely resembles science. It is worth while examining here the roots of the differences between alchemy and natural science in order to discover why alchemy is not simply a rudimentary chemistry, and how far it attempts something that modern science does not.

Modern science, and with it chemistry, makes observations, reports them clearly and without secrecy, extracts from them general laws, explains these in terms of theories, and deduces other laws from these. Moreover, it verifies every step in its inductions and deductions by testing its statements in order to discover how nearly that which has been recorded and inferred corresponds to what is observed, experimentally or otherwise. Natural science admits nothing that cannot be observed, clearly set down, and in some manner verified. It sets itself to interpret the world in terms of a few simple principles, themselves unexplained. Thus we can envisage chemistry as capable of being entirely expressed in terms of a few simple principles, e.g., the electron, proton, neutron, relativity, the quantum law, etc. Yet, however small is the number of principles that may be required, they will still remain unexplained. Here is the first difference between the intention of our science and that of the ancient natural philosophies.

Science deals with that part of the world which is investigable by its methods and makes no attempt to consider the rest. It does not inquire into the ultimate cause of the existence of things; it does not seek to incorporate the world of individual mental activity into its account of what we all perceive in common. The objective of science is limited, and for any one man at any one time, very limited. Each scientist seeks to add a small section to the growing fabric of knowledge, but he does not, as scientist, try to construct a system of the world that includes everything that man can seek to know, past and present.

The alchemist, other than the mere multiplier of metals, sought a complete scheme of things in which God, the angels, man, animals, and the lifeless world all took their place, in which the origin of the world, its purpose, and end were to be clearly visible. Such an object is clearly unattainable by science, for it includes the objects of science, philosophy, and religion. It follows then that the alchemist's attitude and method differed widely from that of the modern scientist.

Alchemy not only sought to deal with matters that chemistry does not approach, but even omitted to inquire into the matters that chemistry has made her own. The alchemists did not seek to establish or even contribute to a descriptive catalogue of chemical substances, nor did they chronicle and classify their changes. We do not find in

232

alchemical works any attempt to list the known kinds of matter and set out their properties, nor yet to make general statements as to the way in which one class of bodies passes into other classes. The eyes of the alchemists were turned on one particular work, the perfection of matter, which in practice meant the elaboration of the stone or quintessence, and this is primarily a *work* to be done and only secondarily *knowledge* to be gained. The knowledge, moreover, that the alchemists sought was not a description of bodies, but a general principle or scheme, in terms of which natural processes became intelligible.

Furthermore, alchemy lacked the close linkage to industry which has always been present in chemistry. Alchemy was certainly intended to be useful. We constantly hear that the alchemist will use his gold to build bridges or churches, to finance crusades or to relieve the poor; or he may seek to use the stone as an elixir to cure diseases. But he never proposes the *public* use of such things, the disclosing of his knowledge for the benefit of man. The alchemist himself will apply the gold or heal the few he chooses to heal. Pearls are not to be cast before swine. Any disclosure of the alchemical secret was felt to be profoundly wrong, and likely to bring immediate punishment from on high. The reason generally given for such secrecy was the probable abuse by wicked men of the power that the alchemical secret would give, and to this reason we cannot but accord some little sympathy. The alchemists, indeed, felt a strong moral responsibility for the result of their work, a responsibility that is not always acknowledged by the scientists of today.

The material aim of the alchemists, the transmutation of metals, has now been realized by science, and the alchemical vessel is the uranium pile. Its success has had precisely the

233

result that the alchemists feared and guarded against, the placing of gigantic power in the hands of those who have not been fitted by spiritual training to receive it. If science, philosophy, and religion had remained associated as they were in alchemy, we might not today be confronted with this fearful problem.

As there was a difference of objective between alchemy and chemistry, so also were there wide differences of method and practice.

The alchemists did not attempt, like the scientists, to rise from particular observations to general rules, from general rules to theories, but obtained their theory from tradition and endeavored to apply it in practice. Alchemy looks backward. The four elements and two vapors of Aristotle, the *pneuma* of the Stoics, and the astrology of Babylon are the origins of their theories. Their research is an attempt to discover the meaning of the men of the past, the men who knew. It was not a great volume of research work that was needed for success in discovery, but the penetration of an individual intelligence.

This difference is reflected in their habits of work, and we can thus understand how it is that science is a co-operative undertaking while alchemy was personal. Each alchemist wished to gain a certain fulfilment by attaining the knowledge of the scheme of things and a mastery over matter. The idea of contributing a little share to man's total knowledge would not have appealed to him; he was doing a work on matter and on himself, and, if he did not bring that work to completion, he failed. We do not find much evidence of collaboration between alchemists. They were more like artists or craftsmen who sought to perfect them-

234

selves in an understanding or wisdom concerning nature which could not be transmitted through written texts. Yet there was one part of the alchemists' equipment which could be handed on, namely laboratory technique, and in this there was a real progress. The Arabian stills are better than those of Mary the Jewess, and those of Brunschwyg and his contemporaries (c. 1500) are better than the Arabian. The alchemists fully realized the fact and merits of a progressive laboratory technique, but they were very sure that this technique was not what made an alchemist. The man who could conduct chemical processes was but one of "Geber's cooks"; the true alchemist was the man who understood the secret ways of things. The artist perfecting his craft and the mystic attaining an inexpressible understanding are more like the alchemist than is the scientist at work on his small assignment, his little share of the building of the stupendous fabric of natural knowledge.

Alchemy again was essentially religious. Its philosophy aimed at the unification of all nature in a single scheme, the author of which was avowed to be God. The attitude of the alchemist towards nature was a religious one. His view was hierarchical; he ranged the substances of which the world was composed in grades of worthiness. The angels were worthier than man; man, than the animals; animals, than plants; plants, than the elements; the fifth element was worthier than the others; fire, than air; air, than water; water, than earth; gold, than the other metals. The changes in nature were thought of as exaltations or degradations in that scale. The alchemist had a direct intuitive appreciation of nature, he reacted affectively to bodies and loved them according to their worthiness, that is their resemblance to the spiritual, the noblest part of nature. This view was made the easier by the alchemist's view of all things as interpenetrated and animated by a living spirit. The world for him was alive and, as Aristotle saw it before him, struggling towards the perfection of God's idea of it.

Science knows simply nothing of all these ideas. Not one thing is worthier than another to the eye of the scientist. If he loves nature, that love is not allowed to enter his books. Matter for him is non-living, and the life he studies in biology is not that life we experience and desire to have more abundantly.

So much then, for the profound differences between alchemy and chemistry; but despite all these the contribution of the alchemists to chemistry is by no means to be ignored. It seems fairly certain that the alchemists invented, and quite certain that they transmitted, the fundamentals of laboratory technique. They taught us how to handle chemicals, how to distil, sublime, filter, and crystallize; they distinguished and named such important reagents as the mineral acids and alcohol. In this respect alchemy is continuous with modern science.

Furthermore, the alchemists founded their work on the idea of natural law. They did not seek to obtain arbitrary or miraculous interventions in the order of nature, as did the type of magician, too common in the Middle Ages, who sought to change the course of nature by the invocation of demons. The alchemist believed that there was a natural process by which gold had been and was being generated in the rocks, and he sought to bring about that process in the laboratory. His theory of the generation of gold was incorrect, but in seeking to do what nature does, he was carrying out what has become a respectable and standard procedure of science. Thus alchemy, in so far as it was a laboratory research based on supposed laws of nature, was on the line of progress that has led to modern science.

Has science today anything to learn from alchemy? Nothing, I believe; for science has been refined until it is an almost perfect instrument for its purpose. No importation of the philosophical or religious into science is possible. But has the *scientist* anything to learn from the alchemist and his medieval contemporaries? Perhaps. He can learn that there are aspects of Nature that do not appear in the scientific journals; that our impressions of her have in them something of man as well as of matter. He can consider her under the aspect of value as well as of disposition in space and time; he can reflect upon the mystery of the existence of the world and of his relation to it.

We shall not return to the alchemists, but doubtless the pendulum, which has swung from the spiritual view of things to the material one, will swing back, and succeeding generations will see the medieval and alchemical concept of nature as a poor foreshadowing of the natural philosophy to which they will have attained.

Recommendations for Further Reading

IT IS NOT EASY to pursue alchemical studies further than the account given in this book without plunging into rather deep water. The list of books here suggested is far from exhaustive, but has the merit of consisting only of reliable works. Many of the older accounts of alchemy are uncritical and lacking in historical sense, and many of the modern ones are mere alchemical tittle-tattle. Unfortunately most of the books here recommended are expensive and rare, and it is very hard to study this subject without either a long purse or access to one of the great libraries.

Among the general works on alchemy which can be commended to a serious reader and which are easily obtainable are:

HOPKINS, ARTHUR JOHN. *Alchemy, child of Greek philosophy.* New York, Columbia University Press, 1934.

JUNG, CARL GUSTAV. *Psychologie und Alchimie.* Zürich, Rascher, 1944. (Psychologische Abhandlungen, vol. V.)

READ, JOHN. *Prelude to chemistry;* an outline of alchemy, its literature and relationships. London, G. Bell and Sons Ltd., 1936.

Much valuable information is contained in the far more comprehensive works:

238

Recommendations for Further Reading

Kopp, Hermann F. M. *Geschichte der Chemie.* Braunschweig, 1843-1847, 4 vols.

Lippmann, Edmund Oskar von. *Entstehung und Ausbreitung der Alchemie,* mit einem Anhange: Zur älteren Geschichte der Metalle. Berlin, J. Springer, 1919-1931, 2 vols.

Journals concerned with the history of science, such as *Ambix,* the Journal of the Society for the Study of Alchemy and Early Chemistry, and *Isis,* an International Review Devoted to the History of Science and Civilization, should, of course, be consulted, also general works on the history of chemistry.

(1) *Works on chemistry before the period of alchemy:*

Bailey, Kenneth Claude. *The older Pliny's chapters on chemical subjects* . . . edited, with translation and notes, by . . . Part 1-2. London, E. Arnold & Co., 1929-1932, 2 vols.

Partington, James Riddick. *Origin and development of applied chemistry.* London [etc.] Longmans, Green and Co. [1935].

Thompson, R. Campbell. *On the chemistry of the ancient Assyrians.* London, Luzac & Co., 1925.

(2) *Alchemy and chemistry of the period 100 B.C. to 1000 A.D.*

The indispensable work is unfortunately very rare, namely:

Berthelot, Pierre Eugène Marcellin. *Collection des anciens alchimistes grecs.* Texte et traduction. Paris, 1888, 3 vols.

Other works of Berthelot which are valuable for this and somewhat later periods are:

Archéologie et histoire des sciences. Paris, 1906.
Introduction à l'étude de la chimie des anciens et du moyen âge. Paris, 1889.

Histoire des sciences: la chimie au moyen âge . . . Paris, 1893, 3 vols.

Les origines de l'alchimie. Paris, 1885.

(3) *Arabic alchemy; the following works are of value:*

HOLMYARD, ERIK JOHN. *The works of Geber,* translated by Richard Russell, 1678. A new edition, with introduction by . . . London, J. M. Dent & Sons, Ltd., 1928.

—— *Avicennae De congelatione et conglutinatione lapidum;* being sections of the Kitâb al-Shifâ. The Latin and Arabic texts; edited with an English translation of the latter and with critical notes, by E. J. Holmyard . . . and D. C. Mandeville . . . Paris, P. Geuthner, 1927.

—— *Kitāb al-'ilm al-muktasab fī zirā 'at adhdhahab;* book of knowledge acquired concerning the cultivation of gold, by Abu 'l-Qāsim Muhammad ibn Ahmad al-'Irāqī. The Arabic text, edited with a translation and introduction. Paris, P. Geuthner, 1923.

KRAUS, PAUL. *Jābir ibn Hayyān,* contribution à l'histoire des idées scientifiques dans l'Islam. Le Caire, Impr. de l'Institut français d'archéologie orientales, 1942-1943, 2 vols.

RUSKA, JULIUS FERDINAND. Numerous works, all in German; especially: *Tabula Smaragdina;* ein Beitrag zur Geschichte der hermetischen Literatur. Heidelberg, C. Winter, 1926.

Turba philosophorum; ein Beitrag zur Geschichte der Alchemie. Berlin, J. Springer, 1931. (*Quellen und Studien zur Geschichte der Naturwissenschaften und der Medizin.* Bd. 1.)

There are numerous valuable papers by H. E. Stapleton, which are, however, not easily accessible, as being contained in the *Memoirs of the Asiatic Society of Bengal.*

(4) *Medieval and later alchemy:*

Much of what has been written on this subject is quite unreliable; the best English source are the chapters on the subject in:

Recommendations for Further Reading

THORNDIKE, LYNN. *History of magic and experimental science.* New York, Columbia University Press, 1923-1941, 6 vols.

Useful articles giving biographical and bibliographical information concerning individual alchemists before 1400 are to be found in:

SARTON, GEORGE. *Introduction to the history of science.* Baltimore, pub. for the Carnegie Institution of Washington by the Williams & Wilkins Company, 1927-1948, 3 vols. in 5.

The texts of this period have for the most part to be read in the original Latin, but A. E. Waite has translated a certain number of these, including:

The Hermetic Museum, restored and enlarged: . . . now first done into English from the Latin original published at Frankfort in the year 1678 . . . [Anon.] London, J. Elliott & Co., 1893, 2 vols.

BONUS OF FERRARA. *New pearl of great price.* A treatise concerning the treasure and most precious stone of the philosopher. English translation. London, James Elliott & Co., 1894.

GRATAROLI, GUGLIELMO. *Turba philosophorum;* or, assembly of the sages, called also the book of truth in the art and the third Pythagorical Synod. An ancient alchemical treatise translated from the Latin, . . . London, G. Redway, 1896.

There are some English alchemical texts of the seventeenth century, but all are rarities:

ASHMOLE, ELIAS. *Theatrum chemicum Britannicum.* London, 1652. (This is a collection of English alchemical poetry, valuable in every sense, and should be reprinted.)

NORTON, THOMAS. *The ordinall of alchemy.* (Reprinted from the above in facsimile, with an introduction by E. J. Holmyard.) London, 1928.

241

The Alchemists

It should be remembered that, since alchemy is a highly obscure subject, translations of alchemical works must contain many doubtful renderings and should not be trusted too far.

Index

243

Index

Index

CPSIA information can be obtained
at www.ICGtesting.com
Printed in the USA
LVHW052318160719
624350LV00001B/11/P